Why Think? ▪

Why Think? :: *Evolution and the Rational Mind*

RONALD DE SOUSA

2007

OXFORD
UNIVERSITY PRESS

Oxford University Press, Inc., publishes works that further
Oxford University's objective of excellence
in research, scholarship, and education.

Oxford   New York
Auckland   Cape Town   Dar es Salaam   Hong Kong   Karachi
Kuala Lumpur   Madrid   Melbourne   Mexico City   Nairobi
New Delhi   Shanghai   Taipei   Toronto

With offices in
Argentina   Austria   Brazil   Chile   Czech Republic   France   Greece
Guatemala   Hungary   Italy   Japan   Poland   Portugal   Singapore
South Korea   Switzerland   Thailand   Turkey   Ukraine   Vietnam

Originally published in French by Presses Universitaires de France as *Evolution et Rationalité* (2004).
Copyright © 2004 Presses Universitaires de France. English translation copyright © 2006
by Ronald de Sousa.

Copyright © 2007 by Oxford University Press, Inc.

Published by Oxford University Press, Inc.
198 Madison Avenue, New York, New York 10016

www.oup.com

Oxford is a registered trademark of Oxford University Press

All rights reserved. No part of this publication may be reproduced,
stored in a retrieval system, or transmitted, in any form or by any means,
electronic, mechanical, photocopying, recording, or otherwise,
without the prior permission of Oxford University Press.

Library of Congress Cataloging-in-Publication Data
De Sousa, Ronald. [Evolution et rationalité. English]
Why think? : evolution and the rational mind / Ronald de Sousa.
p.   cm.
Includes bibliographical references.
ISBN 978-0-19-518985-8
1. Knowledge, Theory of.   2. Evolution.   3. Reasoning.   I. Title.
BD177.D413 2007
128'.33—dc22   2007009407

1 3 5 7 9 8 6 4 2
Printed in the United States of America
on acid-free paper

Preface ⋮⋮

It's fashionable to claim that we should trust our gut, rely on our intuitions, and stop thinking too much. The book now in your hands takes the question seriously: How is explicit human thinking different from the goal-directed "intelligence" of animals? How does our own ability to come to quick, intuitive decisions—often mediated by unreflective emotional responses—relate to reflective thought? The quick responses of intuition often conflict with reflective thought. Yet both have been honed and refined by millions of years of natural selection. So it's important to understand how they both work, and what are their respective strengths.

Evolution itself has displayed a capacity to mimic intelligent planning so uncanny that many people simply refuse to believe it ever happened. That makes it worth asking what natural selection and intelligent thought have in common. Why did the inventive genius of the Wright brothers not enable them to design a 747 straight off the bat? If we ignore the time scale, the path from the Kitty Hawk "Flyer" to the supersonic airliner looks much like the transition from the early *Eohippus* to the modern horse, *Equus Caballus*: gradual, fumbling, step-by-step change, groping forward by trial and error. How then are "rational" solutions different from those arrived at by the mindless processes of natural selection?

This book approaches this question by looking at our nature as rational beings in the light of biology. We don't usually accuse other animals of being irrational, even when their instinctual responses prove fatal: to do so would add insult to injury. When an animal's responses didn't work out, we allow that similar responses must have been good enough in the long run to keep the species alive up till now. But it was all done without discussing alternatives, debating improvements, or mutual criticism. Only humans do all that. We do it mostly by *talking about it*. But our "natural" answers to hypothetical problems, especially those involving evaluation of risk, are notoriously erratic. Human reasoning itself evolved, and took a leap with the invention of language. And language depends (ideally) on informational rather than only on straightforwardly causal processes. Mistakes in reasoning, success or failure, are no longer measured exclusively in terms of biological fitness, in which the only "value" is the reproduction of genes. Why should I care that my genes get replicated? They are not me. I may have other plans.

When canons of rationality can be articulated and debated, disagreement generates a proliferation of individual human values. The crucial transition to deliberation mediated by language is therefore what makes possible, at one stroke, human rationality, irrationality, and the wondrous, chaotic multiplicity of conflicting human values. But where do we get those values? At the deepest level, they have their roots in the very emotions that emerge out of the interplay between our most basic responses in childhood and the elaboration of reasoned ideas, which is what education is all about. Fully to understand this is the goal of psychology and social science. Both must be grounded an understanding of our biological natures. The perspective offered here is therefore a wholly naturalistic one. But if the picture presented in this book has any force, an understanding of the highs and lows of our capacity for rational thought and action can ground a virtually unlimited range of possibilities for human flourishing.

Acknowledgments ⁑

Ruwen Ogien originally encouraged me to set down these ideas, and both he and Pierre Livet provided me with invaluable feedback on the first drafts of this book. I've also learned a great deal from criticisms and comments from Joëlle Proust, Paul Dumouchel, and Frédéric Bouchard. David Egan, playwright, philosopher, and elegant stylist, made many suggestions for improvement and saved me from several inadvertent howlers. Peter Ohlin of Oxford University Press has been a patient guide in the process of making the book more readable, and his encouragement has been invaluable. For the entire period during which I have been working on these ideas, I have been the grateful recipient of Standard Grants from the Social Sciences and Humanities Council of Canada. Last, but not least, I am thankful to my daughter Qingting, not only for daily joy, but for having permitted me to put this book out despite her fear that as a consequence she would no longer be able to claim the status of Middle Child.

Contents ⁜

1 Introduction   3
2 Function and Destiny   29
3 What's the Good of Thinking?   56
4 Rationality, Individual and Collective   87
5 Irrationality   120

Notes   155
References   169
Index   181

Why Think?

Chapter 1 ❏ Introduction

Aristotle called human beings "rational animals." It is all too regrettably obvious, however, that we are frequently irrational. Yet it would be hasty to reject Aristotle's characterization outright. Much of this book is concerned with sorting out how to make sense of both our rationality and our irrationality. It is also about what's good about being rational, and why it's worth the trouble.

To make a start on the latter question, consider Jack and Jill. When Jill tackles a project, she is methodical and scrupulously careful. She tailors her means to her ends. She looks only to the best evidence and the soundest reasons. She is, then, you will agree, as rational as one could be. Yet she fails. Jack, on the contrary, is devoted to Non-Linear Thinking, which he interprets as requiring regular consultation of astrological charts, the hexagrams of the *Yi Jing*, and other magical omens. Rationality, he declares, is over-rated. Irritatingly, he succeeds in what he attempts and loudly trumpets his success.

Such things do happen. When they do, isn't it enough to throw you right off the claims of rationality? What is it, actually, that is so good about rational thinking?

This may seem an idle question. Surely the advantages of thinking are obvious. And yet no product of human ingenuity can hold a candle to the subtle and economical complexity of a single living cell, let alone to the unfathomable organization of what is often termed the most complicated object in the universe, the human brain. Thought nowhere figures in the mechanisms of evolution that have shaped life itself. Nor does it play any part in the procedures used by most organisms to keep themselves alive. Such marvels are not the fruit of any computation or planning: they are merely the upshot of four billion years of natural selection, constrained by the laws of physics, chemistry, and probability. The precise details of the diversity of mechanisms involved in natural selection are still a matter of dispute, but in the main they are adequately summed up in the phrase made famous by the biologist Jacques Monod (1972): *chance* and *necessity*. Nature abounds in astonishing inventions such as the human eye, or the intricacies of the mechanisms that turn food into over three hundred different kinds of cells that make up our bodies. The proponents of the theory of Intelligent Design love to cite these, but they keep having to pick new candidates as science cracks one mystery after another. When a favorite example of the inexplicable is explained, it must be replaced with a new mystery. If the "irreducible complexity" of anything still unexplained had been consistently used to posit the intervention of an Intelligent Designer, evolutionary science would have been abandoned as a waste of time before it started. The wonder of nature's ingenuity rests precisely on the assumption that her most ingenious devices are all natural products of evolution, owing nothing to intelligence. What, then, is the point of thinking?

In approaching such questions, we should first remind ourselves that rationality does not guarantee success. Its advantage consists merely in increasing the *chances* of success. This brings rationality right into line with evolution, of which the very stuff, we might say without much exaggeration, is *probability*. Natural selection has perceptible effects only in the context of large numbers. At the level of statistical phenomena, probability governs the precise interactions of chance and necessity. As for individuals, no matter how well equipped they might be to seize opportunities and face the dangers that threaten them in every natural environment, survival is never guaranteed. What biologists call an organism's *fitness*, its *probable* survival and fertility, guarantees neither its survival nor its fertility. No more, for that matter, than success is assured even in the most minutely planned of intentional undertakings. In both cases, the most we can claim is that the best-adapted organism no less than the most elaborately worked-out plan will be the one most *likely* to succeed. This fact will translate into meaningful observable effects only in the long run, at the statistical level.

In this essential respect, then, the upshot of rational planning elaborated in intelligent thinking is the same as the upshot of natural selection: in the long run, individuals increase the chances of success in their respective undertakings. Furthermore, there is every reason to think that the methods used by rational beings such as we pride ourselves on being have themselves been shaped over millions of years by natural selection. This process took place over an enormous variety of circumstances—when our ancestors lived in the ocean and when they lived on land, when they had to succeed in catching prey and when they had to avoid their own predators. Should we then assume that our strategies of inference and discovery are invariably the best they could be? If not, can we fall back on the thought that they are generally adequate, if not actually optimal? Or should we, on the contrary, resign ourselves to the possibility that the most seemingly "natural" epistemic processes are often ill adapted to the circumstances of present day life? If the more pessimistic hypothesis is right, can we at least still count on our capacity for self-improvement? Given the way we came by our faculty of thought, what reasonable expectations are we entitled to?

The questions I have raised so far are of two kinds. A first batch takes the powers of rational thought for granted. Rational thought is set up as a benchmark, by reference to which we might assess the mechanisms of natural selection. The second batch, by contrast, takes natural selection as its point of departure, in order to question the viability and reliability of our modes of discovery, our rules of inference, and our standards of proof—in short, of all the epistemic strategies that natural selection has empowered us to devise and endorse. Thus the evolutionary point of view suggests two perspectives: one looks at the logic of natural selection that gave us adaptive functions, while the other scrutinizes the origins and the constraints on the rationality of thought and action that supposedly characterizes intelligent human beings. These two perspectives form the framework of what follows.

At the heart of both is the idea of rationality. Let me then begin by attempting to cast a little light on the significance of that notion. Rationality is generally thought to be a good thing, although the occasional dissenting voice is heard to deplore it as rigid, narrow, linear, or even—most horribly—"phallogocentric." What does *rationality* actually mean?

1.1 Two Senses of "Rationality"

At first sight it seems obvious that the ascription of rationality is confined, like its opposite *irrationality*, to thought and action and to organisms capable of both. Talk of rationality is not appropriate in connection with events governed purely by the laws of physics, even if such events involve a rational being. Suppose a man accidentally stumbles and falls into a clump of nettles. We wouldn't label him "irrational," for the incident was not a chosen act. It was a mere event, implicating the person not as an agent but merely as a physical object, subject to the laws of gravitation and inertia. We speak of a falling object as "obeying" the law of gravitation, to be sure, but *disobeying* the law of gravitation isn't really an option. That sort of obedience is neither rational nor irrational.

What this example brings out is that the word *rational* has two senses, marked by two different contraries. In the *categorial* sense, the contrary of *rational* is *arational,* a term that applies to behavior that is due neither to choice nor to thought. The notion of choice, in this context, implies nothing in particular about deliberation or free will, but merely refers to events that are caused in a certain way. For an occurrence to be a matter of choice in the sense intended, its causes must include *reasons*. Reasons, at a first rough level of approximation, provide explanations by appealing to certain goals, norms, or values.

The second, *normative* sense of the word *rational* contrasts with *irrational*. It implies that a belief or behavior was appropriately grounded in specific reasons, norms, or values. In this second sense, an agent who is not rational is in some sense defective in respect of thought or action. Irrationality is a normative notion: its ascription commonly involves a certain sort of reproach, complaint, or criticism. What sort of criticism is a question that will require close scrutiny. For one can criticize a landscape for being dull, or a fruit for being unripe, but complaints of that sort ascribe nothing like irrationality to landscapes or to fruit.

On pain of paradox, the word *rational* cannot be taken in its normative sense in Aristotle's characterization of humans as rational animals. The formula makes perfect sense, however, if it is interpreted in the categorial sense. Which is to say that if human beings can indeed be described as rational animals, it is precisely in virtue of the fact that humans, of all the animals, are the only ones capable of *irrational* thoughts and actions.

The distinction just drawn gives rise to a difficulty, however. If categorial rationality cannot appropriately be ascribed to events that are sufficiently explained in terms of natural laws, does this mean that human behavior escapes the determination of natural laws altogether? One might take this in either of two senses. On a more modest interpretation, it would mean that the laws of physics, chemistry, or any other science—including the laws of probability—that explain the behavior of inanimate objects are insufficient to explain that of rational beings. Rational behavior would then belong in a zone left fallow by the laws of nature and mathematics. Someone might offer the behavior of a chess player as an instance of something that can be explained only in terms of the rules of the game, and rules are not laws of nature. A stronger version would insist that the behavior of rational beings actually transgresses some natural laws. But that thesis would be absurd because to claim that a "transgression" of laws of nature has occurred is to posit a miracle. Or, more reasonably, it amounts to an admission that we hadn't got the alleged laws quite right in the first place.

Some philosophers, such as Kant and Bergson, have clung to the thought that free will transcends the natural world without actually violating the laws of nature. But this attempted solution is bred in bad faith. For talk of transcendence is generally a way of trying to paper over a contradiction with a spot of jargon. Better to acknowledge that regardless of intelligence or rationality, human beings are indeed subject, like everything else, to the laws of nature. The human difference must be sought among natural facts, and not in some hope that natural facts might be transcended.

The evolutionary perspective maintains that life arose about four billions years ago from chemical conditions that are still not fully understood, but of which one can safely presume that they included no phenomena that could be labeled either rational or irrational. It follows that at some point—or perhaps gradually, during a long transitional phase—phenomena that could be classed as rational succeeded others that could not reasonably be so labeled. By similar reasoning, a transition of the same kind must be supposed to take place in the course of development in every individual human organism. For each of us begins life as a single-celled organism, the zygote that results from the fusion of the parental gametes. As that cell and its descendents undergo successive divisions, according to the laws of physics and chemistry that govern those processes, they undergo a series of metamorphoses that at some point gives rise to an organism capable of reasoning, that is, a rational being in the categorial sense.

If we start from the thought that rationality is typically applicable to thought and to action, we can characterize two crucial metamorphoses, both in the evolutionary process and in the course of individual development. One took us from the mere detection of stimuli to the capacity to represent objects; the other took us from tropisms, or automatic behavioral responses, to the capacity to form and act on desires and intentions.

From Detection to Representation

Each living cell, and therefore every multicellular organism, is endowed with some capacity to detect what might be useful or harmful to it. One could call this "sensibility," but the notion I have in mind is meant precisely to contrast with the ideas of consciousness and knowledge evoked by this word. It is better to speak simply of a "detecting function" in order to underline the purely functional character of the faculty in question. The existence of a transition between the detecting function and its rational successor then raises the following questions: At what stage of phylogenetic evolution, and at what stage in the development of each adult to whom rationality is unquestionably ascribed, must we speak no longer of simple detection, but of belief, knowledge, or representations? What are the supplementary capacities that are crucial to this transformation, and how do they arise?

From Tropism to Desire

Every unicellular animal is equipped with a detecting function, on which some specific behaviors depend. In the simplest organisms, this will merely result in approach or avoidance. Although the terms *approach* and *avoidance* may seem to imply a greater measure of mobility than plants can claim, even plants react, if only with a simple change of orientation, the opening of some pores, or the tensing of certain fibers, such as enable the sunflower to track the position of the sun. What counts is that there should be some sort of differentiated behavior corresponding to the information detected. Behaviors of this sort are called *tropisms*, and they are typically triggered by a gradient of temperature, light, chemical concentration, or other stimuli in relation to which the organism orients itself.

Tropisms, like other adapted functions, are the creatures of natural selection. They fulfill tasks essential for the survival of the organism whose goals they serve. Explanation in terms of goals is called *teleological*, so this means that tropisms are liable to be explained in teleological terms. But that word, *teleology*, is rife with potential misunderstandings. When we think of biology in terms of *teleology* and *goals*, are we using these terms in the same sense as when they are used in connection with voluntary decisions and intentional behavior?

If one says of a cell that it *seeks* an environment at a certain temperature, or that it *desires* a certain chemical, one would surely be using these terms in a metaphorical sense. But why are we so sure? What really differentiates a full-fledged desire from a simple tropism? Or to put the question differently, what needs to be added to a tropism to turn it into a desire? This is just another way of posing the question just mentioned: when exactly—on the scale of phylogenetic evolution or on the scale of individual development and as a consequence of what changes in the capacities of an organism—does it become appropriate to speak of desires, projects, and intentions? What precisely make it legitimate to ascribe rationality, and not merely biological functionality, to a given process?

As we have just seen, the categorical notion of rationality implies the possibility of criticism. Three sorts of reproaches, in particular, are appropriate only when they are addressed to a rational agent: it makes sense to criticize a person, but not a cell, for having made a mistake in *computation*, or with having failed to *foresee* what should have been foreseen, or with having acted on *reasons* that fell short of the best set of reasons. We need to ask, then, about the baggage carried by that trio of words: *computation, foresight,* and *reasons*.

This last term is more likely to make trouble than to help. I will pass over it for now, noting only that its kinship with the Latin *ratio* evokes both *proportionality* and *accounting*.

As for the concept of foresight, it seems it could just about be stretched to apply to certain tropisms. A chemical gradient might be said to allow a cell to "foresee," if only in a metaphorical sense sufficient to license prediction of behavior, what it is likely to encounter in one direction or the other. The difference we seek is therefore not likely to be found in the idea of foresight. More likely to be helpful is the consideration that when complaining or criticizing is appropriate, some sort of *norm* must be involved, where a norm is roughly a notion of how things are *supposed* to be. It will therefore be in the neighborhood of this idea of *appropriate criticism* that we are most likely to locate the frontier of normativity, which will allow us to cross into the domain of rationality. We'll have occasion to look into this idea of appropriate criticism. But first, let us look further into the remaining concept in the trio just mentioned: computation.

1.2 What Is Computation?

These days we are used to computers regulating more and more aspects of our lives, so it no longer seems surprising that machines are able to effect computations. But René Descartes (1596–1650), one of the first philosophers who thought seriously about the difference between people and machines, would have been astonished. For he thought of computation as a manifestation of the faculty of reason, and he thought of reason as belonging exclusively in the province of the immaterial soul. It made no sense, Descartes maintained, to attribute the faculty of reason to any sort of material or mechanical device. Some three centuries later, our machines are rather bad at such animal functions as seeing, and they remain awkward in their attempts to get around on two feet. By contrast, they compute all kinds of things with ease, and the best machine is unbeatable at chess, the paradigm of games of intelligent computation.[1] When a machine effects a computation, should we think of it as computing in the very same sense as we might say of chess masters that they compute the next move? This is hotly denied by many champions of the unique human difference. But actually the question glosses over an important distinction between two very different sorts of computing machines: classical *digital* computers and *analog* computers.

To get a sense of that crucial difference, consider what we would think of someone who, after watching Galileo drop stones from the top of the tower of Pisa, offered the following account of the event:

> The stone *computes*, in accordance with the law soon to be formulated by Newton (a formula the stone knows innately), the speed it is to adopt at every instant of its fall.

Simultaneously with this computation, the stone *implements* the motion determined by the result of the computation.

Most of us would assume that this description is meant to be metaphorical or just facetious. When the stone is said to "obey" the law of gravity, that simply means that its trajectory is adequately described by that law. We *use* the law to make the relevant calculations, determining the speed the stone will have reached when it hits the ground. But obviously the stone itself neither computes anything nor executes any plan.

An object that conforms to the law of gravitation can be used, however, to measure or compute something else. A pendulum's behavior is computable from the law of gravity with the help of some geometry and calculus. That provides us with a measure of time, which allows us to "compute" the interval elapsed between two given events. In this way, the pendulum provides a simple example of an analog computer. Another example—though not so obviously useful—is the humble soap bubble: its shape automatically minimizes the surface of a volume of gas, not on the basis of any digital computation, but merely by virtue of the implementation of a physical process.

Similar principles are embodied in self-regulating devices of various sorts. One particularly interesting example is James Watt's governor. This device solves the following problem: how to pinpoint the moment when the speed of a steam engine becomes dangerously high, and slow the engine down to prevent it from racing out of control. Nowadays, we might think of solving this problem by means of a computer equipped with three distinct modules. Call it SPEEDWATCH. A first module would detect the number of revolutions per minute effected by the machine. This module would pass on the information, in digital form, to a second module, which would compare the value acquired with a threshold programmed ahead of time. Once the threshold is reached, a message would be passed on to a third module that controls the pressure in the boiler. That module's task would be to lower the pressure and hence the engine speed. This would be, very roughly, an information-theoretic solution. But Watt's governor has nothing to do with information in any form. It does not detect, compare, or transmit information. Instead it functions purely mechanically.

Watt's governor consists in a revolving central shaft, to which are hinged two wings with weighted tips. As the speed of the machine increases, the central shaft revolves more quickly. The centrifugal

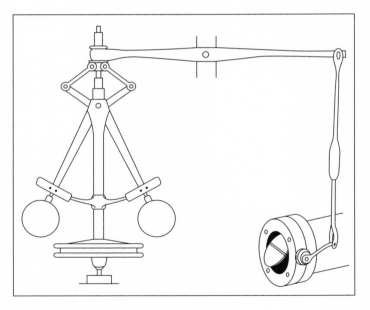

Figure 1.1. Watt's governor

force induced by this rotation lifts the wings, counteracting the pull of gravity that keeps them, when at rest, parallel with the central shaft. As the wings reach up toward the horizontal, they trigger the opening of a valve that allows steam to escape and lowers the pressure. The machine slows down. As the speed eases, the wings come down and the valve closes, which drives the pressure up again. The cycle begins again, keeping the machine close to the target speed.

This efficient and simple device can be thought of as an artificial tropism. Its elegance far surpasses the rather complicated digital-informational alternative described a moment ago. The comparison shows that digital devices are not necessarily superior in all domains, and that computation comes into its own only in certain specific circumstances. We'll have to look into what those circumstances are. The more general lesson to be drawn at this point is that a method appropriate to the solution of some problems at a given scale is not necessarily the best method in all other circumstances and at different scales.

That fact may have set an obstacle in the way of biological evolution. For past a given threshold, a once successful device may block further improvement. Natural selection may then have no means of switching to an entirely different strategy: it would be nice to fly unaided, for example, but we make too much good use of our arms: no amount of natural selection will cause them to morph into wings. That's why prosthetic wings were our only option for emulating birds. (But then those prosthetic wings easily outstripped the natural ones.)

Does that mean rationally designed technology can overcome any problem? No: rational deliberation can face a similar difficulty. In many practical circumstances it would be counterproductive to embark on minutely detailed planning. When a quick reaction in real time is imperative, an approximate but fast response will be more effective in the long run than an exact solution that takes time to work out. By the same token, however, the quick response may lead us astray. This is an inescapable dilemma of rationality. As we shall see, it has many interesting consequences.

Watt's governor is an analog machine. Strictly speaking, it performs no computations. Its functioning rests merely on physical laws that govern all dynamic systems—whatever involves forces, velocities, and masses, the relations among which are summed up in Newton's formula "Force = Mass × Acceleration." In a way, the calculating machine or computer does nothing different. It, too, has been conceived and built to behave in a certain way in the light of the laws of electricity and the logic of its circuitry. And yet surely the distinction between the two does not depend merely on the difference between dynamics and electricity. So what is the essential difference?

1.3 Digital and Analog

The best way to sum up the nature and the advantages of digital representation is to recall the paradoxical theory of resemblance advocated by Plato, known as the *Theory of Forms*. The best illustration of the core idea is provided by our humble roman alphabet. Take two tokens that resemble one another, say a and *a*.[2] Is the resemblance between them a two-term relation? Obviously yes, says common sense. But Plato held that it should really be treated as a three-term relation: what the resemblance between a and *a* consists in, he proposed, is the fact that both *derive* from a third entity, call it A, which is the "ideal" A and which is the depository of the essential nature of both a and *a*. There are, in fact, a thousand different fonts and styles of type, of which the tokens of 'a' resemble one another less than they might resemble other letters.[3] What all instances of the letter a have in common is just that they meet whatever norms it is that constitute them as concrete instances of that particular vowel. In practice, such norms are often ill conceived, and so we are apt to use contextual clues to avoid confusing the number 1, the uppercase ninth letter of the alphabet, I, and the lowercase twelfth letter, l. But in a well-planned and designed digital system, the elements of the system would be sufficiently well spaced as to make mistakes virtually impossible.

As everyone knows, computers of the sort many of us now use every day are based on a system of digital representation. What that means is that the voltage changes effected in the computer's circuitry are regarded, for the purposes of computation, as taking only one or another of a finite predetermined number of discrete values. From the purely physical point of view, of course, the voltage changes are effectively continuous. This illustrates the importance of distinguishing between the characterization of the physical processes themselves, and the way those characteristics are interpreted when they are set up to be part of a digital representational system.

Watt's engine governor shows that a mechanical or dynamic system, in which all effects derive simply from behavior that conforms to the laws of nature, can provide the most effective solution to certain classes of problems. By contrast, a device based on digital representation comes into its own whenever there is a need for a great many faithful reproductions of a given original. The reason is that all dynamical systems are subject to small variations and generate small errors of measurement at every stage. Whenever a sequence of copies is required, those tiny initial errors will cumulate and turn into serious discrepancies, as in successive xeroxes of xeroxes that quickly become illegible. In a digital system of representation, by contrast, you are never further than two steps away from the original. For in a digital setup, as in

Plato's theory, you don't really copy a copy, or indeed even the original. Instead, you copy the "idea," the paradigm, of which the original was itself only an instance or copy. Each copy derives, not from the product of previous copying operations, but from the original itself. Digital systems of representation and reproduction afford *almost perfect* copying fidelity in arbitrarily long series.

DNA, the digital "language" of genes, affords a remarkable illustration. Some sequences have remained almost the same for over a billion years. Some of our own most important genes, such as the "Hox genes" governing the basic body plan of bilateral creatures, have remained almost unchanged for some two billion years. They first operated in organisms that are the ancestors both of humans and of insects (Carroll 2005). (The *almost* is essential, however, for without the occasional copying error, we would all be single-celled organisms, like our very distant cousin the amoeba. Absolutely perfect copying would have allowed no mutations, and therefore no evolution.)

The degree of fidelity with which organisms reproduce is due in large part to the digital "language" of the genes, which guarantees an infinitesimally small mutation rate.[4] The analogies with actual language are striking. First, we can speak of an *alphabet*, constituted by the four bases of DNA or RNA—cytosine, thymine (or uracil in RNA), guanine, and adenine. Taken three at a time, these yield possible combinations that suffice to constitute enough *words* to specify each of the twenty-odd amino acids of which all proteins are made; those words, in turn, link into *sentences* that determine the immense variety of possible proteins. Finally, by analogy with *discourse* built up of sentences, we can think of the proteins making up all the works of biological nature; that is, of all living organisms, as well as the enzymes that are essential to the elaboration and differentiation of their organs.

1.4 Individuals and Communities

The elaboration of a mature organism, partly on the basis of the information provided by its DNA in the presence of favorable environmental conditions, requires a complex and precise collaboration between the individual cells that make it up. This collaboration shouldn't be taken for granted. How did some organisms come to diverge from the condition of their remote ancestors, living as independent individual cells fending for themselves, to arrive at the condition of such complex differentiated multicellular organisms? However it happened, that was assuredly one of the "major transitions" of evolution (Buss 1987; Maynard Smith and Szathmáry 1999).

This observation serves to remind us of another contrast, which cuts across that drawn above between the effects of natural selection and those of purposeful action. This is the contrast between individual rationality and group rationality. It often seems natural enough to apply the concepts of rationality and irrationality to groups or societies. Indeed, some philosophers have spoken of "collective selves" or "plural subjects," capable of formulating and executing genuine group intentions (Gilbert 1992). But as we shall see, the relation between individual rationality and group rationality raises difficult problems, no less at the theoretical than at the practical level.

Individual rationality seems more straightforward; yet we can also think of a single individual as a sort of community. Plato claimed that the soul has three parts, often at odds with one another. Freud favored a rather similar metaphor and distinguished ego, id, and superego. And at the level of the body of any multicellular organism, the viability of the whole presupposes a complex cooperative system linking the myriad cells of which it is composed. From that perspective, cancer can be thought of as the consequence of a break in that collaboration: the refusal, by a certain lineage of cells, to continue subordinating themselves to a coherent whole, for the sake of a proliferation that benefits only itself.

1.5 The Origins and Limits of Intelligence

At first sight, the difference between the products of intelligent thought and those of natural selection seems obvious enough. Thought boasts a crucial differentia. When pressed to declare what this is, some philosophers favor *intentionality*, others, *consciousness*. But most would accept that the differentia, however it should be characterized, amounts to the presence of *mind* or the *mental*. But what precisely is the mental?

To consider answering this question in a naturalistic framework is to entrust the analysis of the evolution of mind, as a product of natural selection, to mind itself. Should we worry that this makes mind into both judge and party? That might raise a doubt about the reliability of the naturalist perspective. Such is the charge made by the theist philosopher Alvin Plantinga.[5] His challenge can be paraphrased as follows:

> If indeed mind has its origin in natural selection, we have no reason to believe that our mental faculties are capable of uncovering the true nature either of mind itself or of the process of natural selection that allegedly gave rise to it. For it evolved under circumstances that couldn't possibly reward the ability to reveal such truths.

The ingenuity of this challenge lies in its attempt to use the very tools of naturalism against itself. It impugns not natural selection as such but the compatibility of natural selection with our capacity to know about it on the basis of rational scientific inference.

In reply, one might first point out that the theist alternative Plantinga touts has its own problems. For let us suppose we knew that our brain was fashioned by some creating intelligence rather than by natural selection. Nothing of interest could be inferred from that, unless we knew the *intentions* with which the creator fashioned it. But the theological premises required to deduce that the creator intended to favor us with a faculty for blanket truth-discovery would be entirely arbitrary.[6] Long before the hypothesis of natural selection became available to provide a concrete alternative, David Hume ([1779] 1947) had already perceived that the inevitable uncertainty about the putative creator's intentions suffices to knock theism out of the field of serious contending hypotheses.

Nevertheless, Plantinga's challenge remains all the more intriguing if it is regarded independently of the theological leap he urges us to make. Stripped of theology, we can reformulate it thus:

> The circumstances under which natural selection honed our mental equipment were presumably such as to afford our ancestors solutions to the practical problems of survival they confronted. We have no reason for thinking this same equipment is epistemologically reliable in situations that differ materially from those faced by our distant ancestors.

Actually Descartes had already come up with essentially the same worry over three centuries ago. In the fourth *Meditation*, he warned that we risk falling into error if we insist on inferring significant propositions about objective reality on the basis of the information provided by our senses (Descartes [1641] 1986). The only thing of which we may be reasonably certain (given God's benevolence) is that our senses will provide sound guidance for getting around in the practical world. But nothing can be inferred about the actual nature of the world itself.

Plantinga's challenge differs only in one particular from his predecessor's warning. For Descartes, God's goodness warrants two things. First, it allows us to count on the utility of the senses, but not on the truth of any ontological inference we might make from their deliverances. Second, it warrants the correctness of whatever we conceive clearly and distinctly. For Plantinga, even in my theology-free version, evolution can take over the first warranty but is very far from being able to underwrite the second. We are therefore in a weaker epistemological position than Descartes imagined. For we have no good reason to believe that what seems clearly obvious to us is actually true. Consequently, we can do no better in our quest for truth than to trust what seems to us most probable after we have looked into a question as thoroughly as possible. In any given case, we must admit the possibility that we might be mistaken. It is therefore crucial that we respond to Plantinga's challenge. If we cannot, we will remain massively vulnerable in all domains in which our reason ventures beyond the observable facts. We need to set out the reasons we have for trusting the methods of rational thought that evolution has bequeathed us.

1.6 How to Meet Plantinga's Challenge

Plantinga speaks to a view that is currently popular with a great many epistemologists, under the name of "reliabilism." Instead of searching for unshakable foundations of knowledge and rationality, reliabilists are merely committed to the idea that natural selection has so tuned our faculties as to make it likely that our search for truth will have a happy outcome in most practical and scientific contexts. More specifically, some have speculated that the reliability of reason is the fruit of an interaction between natural selection, which has shaped our individual intellectual capacities, and the effects of social intercourse, particularly as enhanced by the invention of language. On this view, the confidence we have in the opinions of others would play an ambiguous role, somewhat analogous to that played by our predisposition to make inferences from our experience about what lies beyond it. Induction—as such inferences are called—generates superstitions as well as correct predictions. But one cannot hope to avoid superstition without at the same time increasing the risk of missing certain significant correlations. Similarly, our tendency to believe what others tell us can be harmful as well as useful, depending on the circumstances. But if one were to try to protect oneself against the bad effects of credulity, one would immediately deprive oneself of the great advantages of submitting to mutual criticism. We need to tread a fine line: too much trust can be as harmful as too little. As we will see in chapter 4, it would be rash to expect natural selection to have set up a social contract that guarantees truth-telling. And in chapter 5, I shall explain how, although true belief is not the inevitable consequence of social conformity, conformism might nevertheless have secured the minimum of homogeneity to allow for some kind of group selection, which may at least sometimes favor strongly cooperative practices.

Despite its plausibility, I do not believe that reliabilism provides an adequate response to Plantinga's challenge. Without claiming a high degree of certainty, a slightly more specific hypothesis seems to me more promising. This hypothesis looks to mathematics, regarded as an extension of ordinary language. The argument rests on two propositions: (a) mathematics is unique from a methodological point of view; and (b) its usefulness to scientific and technological progress provides an independent argument in favor of its claim to present us with objectively correct representations of an external world.

Let me expand on each of these two points.

(a) On the methodological front, it is obvious that the development of our brain could not possibly have been influenced by selection pressure specifically favoring the faculty to do pure mathematics. Natural selection could not have gotten a grip on the ability to do mathematics as such unless that capacity manifested itself in some individuals while failing to do so in others. But mathematics is a recent invention: as far as we know, no one did pure mathematics until long after the human brain had attained essentially its present size and capacities, no more than a few thousand years ago. On the evolutionary scale, mathematics is part of our present rather than of our evolutionary past. It is therefore out of the question for mathematical talent as such to have been a factor in evolution by natural selection. Unless it is a faculty that somehow altogether escapes the constraints of biology, the mathematical faculty had to develop as a mere side effect of some other, more immediately useful, set of intellectual capacities in our ancestors. The mathematical gift remained long in the shadows, as it were, behind more visible talents, until the time it emerged fully mature, like Athena out of the head of Zeus.

(b) Once mathematics had emerged into the light of day, there was still nothing to guarantee that it could prove useful outside the domains in which our practical skills had already been operating for millennia. And yet, pure mathematics notoriously finds all kinds of startling applications in the solution of technological and scientific problems that our ancestors could not possibly have conceived of, and it does so by generating theories that would have remained wholly unintelligible to them. That strongly supports the idea that mathematics can uncover aspects of the universe of which neither the usefulness nor even the existence could possibly have been manifested in the environment of our evolutionary adaptation (EEA) in which the basic functions of the brain were being shaped by natural selection. As Wigner (1960) has argued, this constitutes at least prima facie evidence for the conclusion that the truths of mathematics do not merely reflect projective constructions of our brains, but probably correspond to an objective reality. This fact remains deeply puzzling, to be sure, since it can't have been a direct consequence of any sort of tuning of our brains to the world by natural selection. But the mystery is scarcely likely to be cleared up by being ascribed to divine intervention. The theological hypothesis is no real alternative. The upshot is that we can pursue the present inquiry without fear of its being reduced to absurdity either by Plantinga's attempt to show that the belief in natural selection undermines itself, or in virtue of the trickery of some evil demon.

Despite this methodological reassurance, there are compelling reasons to regard our thought processes as leaving much to be desired. On the basis of apparently solid premises, using unimpeachable rules of inference, we can be led to insoluble paradoxes, some of which have kept philosophers busy for centuries. How, asked Zeno of Elea, can I get from A to B in a finite time, since I must get halfway, then halfway again, an infinite number of times? If I assert: "What I am now saying is false," have I lied? Some types of problems, particularly those in which we are required to reason about probability, are liable to lure even individuals trained in the art of careful thinking into drawing systematically erroneous conclusions. Such pathologies of reasoning challenge us to inquire into the causes of such mistakes. In addition, they invite even more disconcerting questions: how could it be that natural selection should have shaped our minds in such a way that we are systematically prone to error? Does this supposition not contradict what we assume we know about natural selection, namely, that it leads to adaptation and thus optimizes fitness by perfecting the organs and capacities on which it works? Or must we fall back on the exculpatory supposition that, as a relatively young species, our faculties have still a long way to go before they reach the perfected state for which they are headed? Shall we need another few hundred thousand years for the kinks in our mental powers to get ironed out? That would be an intellectually flaccid way to explain away the mental deficiencies that have been observed in our species: let's hope for a more interesting approach.

1.7 Prospect

Adaptation, of organs to their function and of organisms to their ecological niches, is a particularly striking feature of the living world. That observation, however, often encourages a hasty inference to the conclusion that teleology is ineliminable from biology. It can seem undeniable that in order to explain the behavior of every organ and every member of a biological community—organism, hive, or ecological web—we must refer to the teleological aspects of its organization.

Chapter 2 will be devoted to the examination of that presumption of teleology. I shall argue that although teleology is not entirely banned from biology, it subsists only in what I shall refer to as a "vestigial" form. In that vestigial or degenerate form, it is still capable of explaining why nature produces such a powerful impression of being pervaded by inherent teleology. But teleology in that form is wholly distinct from that which is in question when we speak of the goals, purposes, and values of individual human agents.

In chapter 3, I shall return in more detail to the resemblances and differences between the "methods" of natural selection and those of reflective thought. These comparisons will lead us to consider the difference between digital and analog systems of representation. I shall argue that the distinctive characteristics of human thought rest essentially on the digital features of language, and that the novel capacity for explicit thought afforded by language enables the proliferation of individual values. This proliferation of values not only detaches human goals entirely from the vestigial "goals" of nature but also generates conflicts, both within and among human individuals as well as communities.

In chapter 4, I shall further explore the relation between individual and collective rationality. A sketch of some of the better-known models of emergent collective organization on the basis of purely individual interactions will serve to illustrate some of the advantages, problems, and limitations of the move from individual to collective rationality. That will bring me to one of the more heated controversies in recent theoretical biology. That is the debate concerning the units of selection, group selection, and the main explanations that have so far been adduced to explain the paradox of individual altruism: how is the Darwinian hypothesis of a universal "struggle for survival" compatible with the fact that human beings sometimes actually sacrifice themselves for a cause that benefits only others?

In chapter 5, my central protagonists—evolution and rationality—will flip over and trade places. Having looked, in previous chapters, at the aspects of natural selection that seem most akin to intelligent planning, I shall turn to the evolution of our rational faculty for intelligent planning. I will bring forward a sampling of some of the examples of systematic irrationality that have been brought to light by psychological research and will explain the apparent paradox that the much-vaunted human differentia of rationality, in the categorial sense, is actually grounded in our capacity to manifest irrational thought and action. I will show how even the most extreme case of irrationality presupposes a minimal level of normative rationality. We shall also see how the most notorious cases of irrationality can be explained on the basis of three factors: the modular organization of our capacities; the widening gap between the vestigial "goals" of natural selection and the concrete goals that human agents set up for themselves; and the indispensable yet ambiguous role played by emotions in the economy of rationality. Although our cognitive predispositions do entail serious deficiencies in a number of types of concrete situations, that is no reason to expect that future evolution will put this right by bringing a higher level of perfection to our cognitive powers. For in the great majority of cases, our faculties' defects are just the flip side of their virtues.

Chapter 2 ⁜ Function and Destiny

Every religion, practically every philosophy, and even some of science, all bear witness to the tireless, heroic and desperate effort of humanity to negate the contingency of its own existence.

—Jacques Monod

Two characteristics of the living world are particularly striking: diversity and adaptation. The diversity of living things is so great that it seems practically impossible to conceive of a novel science fiction creature that does not resemble something already to be found in nature.[1] From bacteria to dinosaurs, from protists to roses, and from prions to insects, the profusion of forms found in nature appears to be the product of a prodigious imagination. An inventive (though somewhat erratic) genius also seems manifested by the way every organ is adapted to its function. As one philosopher of biology has remarked, this fact has left its mark in the vocabulary of anatomy, which is full of metaphors and analogies drawn from technology. Witness, for example, the words: "trochlea ['pulley']; thyroid ['shield-shaped']; scaphoid or navicular ['boat-shaped']; hammer; sac; tube; thorax ['breastplate']; tissue; cell, etc., ... [all of which terms] assimilate parts of the organism to tools or functional parts of machines" (Canguilhem [1966] 1994, 323). It is not surprising, therefore, that we should be tempted to attribute such complex and protean adaptations to a creative intelligence.

2.1 Cause and Teleology

Anyone inclined to yield to that temptation will favor explanations of natural facts modeled on the commonsense explanation of ordinary human action. This requires both efficient causation, referring to forces and mechanisms that make things happen, and teleology, which is concerned with the purpose, goal, or function *for the sake of which* things happen.

Suppose I light a lamp. Causation is involved twice over: first, because we must suppose that some intention of mine was the *cause* of my undertaking my lamp-lighting project. Second, because beyond the "basic act" performed—the pressure of my finger on the switch—every successful action relies on a chain of causal consequences of that basic act, culminating—if all goes well—in the result intended. The pressure must act on the switch; the switch must close the circuit; and the circuit must be live and light up the filament.

But teleology is also involved, since the very possibility of the plan and its implementation depends on the existence of an intention. If the pressure of my finger on the switch was purely accidental, or occurred against my will, the resulting event is not something I am fully responsible for: in a strict sense of the word, it is not an "action" at all. When I do something, teleology is involved regardless of the nature of the act, which can be something as simple as lifting a glass to my lips or as complicated as building a computer.

Some philosophers have claimed that an explanation cannot invoke both causation and goal-directedness. The reason sometimes adduced is that the link between the intention and the act is not causal, but conceptual.[2] There is in this view a grain of truth buried inside a mistake. What is true is that any action worthy of the name must be capable of being judged more or less rational. But while rationality is not *merely* a matter of causality, wherever there is intentional action, rational or irrational, there must *also* be causality. For an act to be rational, it is not enough that there should be a reason for it. It must also have been done *because* of that reason.

I stressed in the last chapter that the normative sense of rationality presupposes the categorial sense. I should now add that the converse also holds. To see why, consider the tragic case of Andrea Yates, condemned to life imprisonment in Texas in 2002 for having drowned her five children in the bathtub. The defense did not dispute the facts, but had pleaded not guilty on the grounds of insanity. Andrea Yates had acted under the influence of voices from God, commanding her to drown her children in order to save them from Satan. This defense was rejected because she had executed her project with deliberate method, proving that she was fully aware of what she was doing. Since she had acted *rationally* in the light of the goal she had set herself, the jury inferred, she could not have been insane.

The verdict is shocking because it takes account only of the minimally rational structure of Yates's action. She had indeed proceeded methodically in the undertaking of drowning her children. In relation to that restricted framework, therefore, her act was indeed rational. But as soon as we ask about the justification of the undertaking itself, the judgment of rationality must obviously be reversed. In relation to a wider framework, it is the project itself that is insane. Can we hope to define an objective, absolutely valid framework in the context of which one could make a definitive judgment about rationality? Well, think of Agamemnon and of Abraham. These two men have not universally passed for criminal or insane, though both were bent on killing their own child. Agamemnon actually did it; Abraham refrained at the last minute.

Abraham, indeed, formed that intention just like Andrea Yates: on the basis of hearing instructions from God. That suggests that Andrea Yates's particular bad luck in being found guilty of murder was in part due to the fact that the psychosis from which she suffered was merely an individual, not a collective one. When enough people share a delusion, it loses its status as a psychosis and gets a religious tax exemption instead. In the light of that reminder, Yates might have lived in a context in which an anthropologist, properly mindful of cultural relativities, would have absolved her of any insanity, just like Abraham or Agamemnon.

However that may be, here is the moral to be drawn from the sad story of Andrea Yates. Any judgment of rationality or of irrationality depends on a given framework of relevant considerations. In the context of the most restricted framework, all action is by definition rational. This basic sense of rationality is admittedly trivial, even tautological, since it follows simply from the fact that the agent does something "because she wants to." That is the sense in which categorial rationality, even in cases of the most egregious irrationality, presupposes a minimal level of normative rationality.[3]

This sketch of the role of goal-directedness in action is obviously simplistic. It takes no account of the subtle distinctions to be made among acting deliberately, intentionally, with previous intention, consciously, willingly, willfully, against one's will, out of weakness of will, and so forth.[4] But it suffices for my purposes here, where it is meant only to prepare the ground for an introduction to the role of teleology in biology, in which intentions have no role to play.

Teleology and Fatalism

Causes precede their effects. Teleological explanations, by contrast, appeal, not to what preceded the event to be explained, but to its *end*, in both senses of that word. An end is both the conclusion of a series of events—their terminus in time—and also the *goal* toward which the preceding stages of the processes tended. At first sight, then, teleology looks like a sort of backward determinism, in which the explanation of a series of events is sought in the future instead of the past. If that is how teleology is conceived, however, it is more like fatalism than determinism. The core of the notion of determinism is this: *Every event is the specific effect of antecedent conditions jointly sufficient to produce that event according to natural laws.* If determinism is true, then, changing any one of an event's antecedent conditions may result in the nonoccurrence of the event. By contrast, a teleological explanation is not concerned with antecedents or their consequences. Instead, it views an event that has not yet occurred as *bound* to take place, regardless of antecedents. It was fate, not causality, that decreed that Oedipus would kill his father and marry his mother. If it had been a question of interfering with a deterministic chain of causation, a single difference in the sequence leading up to the fated finale should have sufficed to derail Oedipus' destiny. But in his case, on the contrary, his every effort to elude his fate inevitably turned into the means of its fulfillment.

At the limit, determinism and fatalism would coincide. Let's suppose with Laplace (1830) that we could know with absolute precision the properties of every particle in the universe.[5] If that were so, and if the laws of nature were fully deterministic, they would allow us "in principle" to deduce the entire evolution of the universe. Time would be reversible. Apart from the detail that we "move forward" in just one temporal direction, the future and the past would be perfectly symmetrical, and it would be just as logical to appeal to what must "fatally" happen in the future to explain the past, as to explain the future on the basis of past determinants.

The view of the universe that Laplacian determinism set itself up against was that of Aristotle. Aristotle distinguished the notion of "final cause" from that of "efficient cause." The latter most nearly corresponds to commonsense causality, or to the notion of cause that figures in the modern doctrine of determinism. But for Aristotle, nothing could be fully explained in terms of efficient cause. Any complete explanation required, in addition, a knowledge of the determining roles of *matter*, of *form* or essence, and of the *end* or intrinsic teleology of the phenomenon in question. This end represents the realization of any natural thing's *potentiality*. Thus, says Aristotle, a man, the child's father, is the efficient cause of the child's birth because he is involved in the event that starts the process of gestation. He is also a part of the material cause, in providing the semen. And "man" is the formal and final cause as well because becoming a man is what the baby is supposed to do.

It might seem that on this view everything is determined in a fatalistic way. Yet pure teleology is not fate. What is fated to happen will necessarily happen. For Aristotle, by contrast, natural teleology is manifested not invariably but "always or for the most part." Unlike fate, a natural end may remain forever unactualized.

In sum, Aristotle's world differs from Laplace's in two ways. First, in that it contains teleology; but second, in that it is not ruled by uniform necessity: any given potentiality may fail to become actualized.

This looks more like the world we know than does the vision of Laplace. But it does not go far enough. For in the real world of living things, it is all too evident that failure is no less frequent than success in the realization of potentialities. In regularly letting us down, our organs are the lifeblood of a vast medical industry that would quickly collapse if our bodies functioned properly "always or for the most part." On the phylogenetic scale, biologists commonly agree that at least ninety-nine percent of all species that ever lived are extinct. (This problem did not exist for Aristotle, at least on the received view that attributes to him a doctrine of the immutability of species.) It is true, as the Gospel of John [12:24] observes, that "except a corn of wheat fall into the ground and die, it abideth alone: but if it die, it bringeth forth much fruit." But John presumably meant that the seed's death is necessary for it to bear fruit, not that it is sufficient. He failed to add that the immense majority of seeds die without bearing anything at all. There is no getting around the fact that wherever it seems plausible to posit a teleological end, that end may never be reached. That poses a major problem for the interpretation of teleology as backward causation. For it is implausible enough to put the cause after the effect; but it is downright absurd to attribute causal powers to what fails ever to occur at all.

Cause and Causal Explanation

It seems intuitively plausible to explain things in terms of what they are for. That is one reason it has seemed tempting to assimilate teleology to a form of causation. Even if one assumes that all scientific explanation must be causal, however, the notion of cause itself does not coincide with the notion of *causal explanation*. The reason is that the cause-effect relation holds between particular events, regardless of the description under which those events are picked out. By contrast, an explanation is required to make the relation *intelligible*. Suppose, for example, that you tell me that what happened at precisely 8:00 AM caused an event that took place at 8:00:01. I may be quite ready to take your word for it, but that doesn't mean I now have an explanation. For that to be the case, I need to be able to class the two events involved under descriptions that figure in some law or generalization. It might be something very informal: what happened at eight, for instance, was that I jogged a vase on the edge of a table and thereby displaced it. From the fact that the vase in question was deprived of its support, in conjunction with the law of gravitation, I can then deduce that the vase will have fallen to the ground one second later. This commonsense explanation illustrates the fact that certain descriptions are privileged by the demands of explanation: an effect cannot be explained in terms of its cause unless both are described in the appropriate way.[6]

Furthermore, causal explanation is not the only form of explanation there is. The reader will recall that Aristotle had stipulated that in order fully to understand a phenomenon we need not only to know its efficient cause and its end or *final cause*, but also its *matter* and its *form* or essential property. Despite its quaint old-fashioned flavor, this allocation of explanatory tasks can still be quite instructive. Let us see, to begin with, what might be gotten out of the idea of *material explanation*. Until 1828—the date of Friedrich Wöhler's successful synthesis of urea—it was generally assumed that all living things were constituted of a specific "organic" matter, irreducible to any kind of matter that made up the inorganic world. The very nature of that special kind of matter guaranteed that it would organize itself in such a way as to serve the necessary vital functions of an organism. This was roughly the import of the doctrine of *vitalism*. That view might be summed up, in a parody of the slogan attributed to Pierre Jean Georges Cabanis (1757–1808) that "the brain secretes thought as the liver secretes bile," as the doctrine that *organic matter secretes teleology*.

We know now that the characteristics of living beings are not to be explained in terms of properties of exclusively organic elements. Nevertheless, a certain number of elements—carbon, nitrogen, oxygen, and hydrogen—are widely found in the biosphere. It might therefore seem plausible that living things owe some of their properties to the matter of which they are typically made. Some thinkers, notably Stuart Kauffman, have propounded the hypothesis that certain sorts of complex systems have an inherent capacity for *self-organization*. They spontaneously arrange the relations of their parts in such a way as to provide a certain measure of "*order for free*," where that phrase implies that the order does not result from the intervention of natural selection but on the contrary provides a relatively limited range of ready-made types of ordered systems among which nature gets to select. This is suggestive: one might read into it a kind of abstract sense of material cause, stemming not from the intrinsic properties of this or that kind of matter but from the potential of any sort of matter for meeting certain conditions for self-organization.[7]

This idea of Kauffman's, whatever its other merits, serves at least to remind us that order is not synonymous with teleology. There is a huge diversity of highly organized forms to be met with, for example, among complex snowflakes, to the extent that their diversity rivals that of living forms (Libbrecht and Rasmussen 2003); but that doesn't mean that snowflakes aim at any goal or that their forms should be explained teleologically. Nor is order synonymous with improbability. Suppose you pull a sequence of numbers out of a hat. The chances of getting any particular random series of digits is no greater than that of getting any particular series that appears highly ordered, such as one that regularly repeats "0123456789."[8]

There are, in sum, three ideas that tend to be confused but need to be distinguished: *improbability, organization*, and *teleology*. The first two are often an indication that it is worth looking for the third, but they do not entail it.

Consider improbability. People commonly marvel at the improbability of all kinds of events, ranging from chance encounters to the creation of the universe. From that claim of improbability, they fallaciously infer that we must seek some explanation in terms of providence or of the intervention of some intelligent creator who alone could plausibly be supposed to have brought about such an unlikely event. The Polish writer Stanislaw Lem has wittily shown the error of that reasoning by pointing out that the existence of any particular human individual can be assigned a prior probability so infinitesimally low that it easily meets the most demanding criterion anyone might offer for saying that something is *impossible*.[9] The logic behind the principle of inference that goes from *improbability* to *providence* should then force me to conclude that *I do not exist*. For my existence required the chance encounter of one among hundreds of millions of spermatozoa with one among hundreds of thousands of ova. Each cell involved in that lottery was itself the product of chance events of staggering improbability, in a chain that stretches for billions of years into the past. Once all these improbabilities are multiplied together, the chance of my existing are astronomically smaller than the chance that I might hold the winning ticket in a lottery with as many tickets as there are elementary particles in the universe.[10]

And yet, I exist. The probability of any phenomenon that has actually occurred is 1.0: that my own birth happened is now absolutely certain. Like any other event that marked the beginning of the life of an individual or of a species, my birth was a unique event of which the intrinsic probability was infinitesimal in relation to some sufficiently distant prior situation. When this is added to the systematic unpredictability resulting from the influence of chaotic and stochastic or purely chance phenomena, this observation should scotch the temptation to believe that whatever is both highly improbable and highly ordered must be due to some creative intelligence.

But if not a creative intelligence, what else could be the cause of order? Consider snowflakes again. As far as they are concerned, the answer to the question just posed is the Aristotelian one, that matter contributes to the complete explanation of any given phenomenon. In this case, the answer has to do with the molecular structure of water, which explains the hexagonal structure of snowflakes, while the vicissitudes through which flakes pass as they encounter changing conditions account for their diversity.

The structure of H_2O, in fact, is perhaps better classed as a fact about the *form* of water, in the literal sense of *shape*. But the distinction between form and matter is not an absolute one. Possible shapes for any given lump of matter depend on the kind of stuff it is. But in turn the characteristics of the stuff depends on the form of its basic elements. Similarly, the capacities of living things clearly depend, in part, on the stuff of which they are made. Should we infer, then, that no living beings could exist unless they were made of just that stuff? Could robots made of silicone be living things? Every sort of matter has its own peculiar capacities, and the complexity of the living world is such that the laws of chemistry might not allow any other combination of elements to replace those that currently constitute our biosphere, and specifically all those parts of it that are constituted of proteins. In the light of that possibility, Aristotle seems to have been right in claiming that the matter of which a thing is made contributes to the explanation of its properties. Aristotle also claimed that a satisfactory explanation must refer to an *essence*. Ignoring scholarly debates about the exact meaning of that word for Aristotle, the following simple definition will do for my purposes here: *The essence of a thing X is the set of properties in the absence of which X would cease to exist.* The essence of the snowflake derives from the structure of its atoms, but it will cease to be a flake, without changing its matter, the moment it melts and loses its specific form. The Aristotelian notion of form is not limited to physical shape, since it refers to the whole set of properties that are essential to a thing, that is, those properties that are responsible for its being the very thing it is. (The Greek phrase used by Aristotle is quite striking: τὸ τί ἦν εἶναι means, roughly, *what it is to be this*.) In the case of any artifact, what it is to be it is typically to be intended for some specific function. In the case of that sort of object, to have a nature and to have a function are one and the same. What is a chair? It's a piece of furniture that is intended to be sat on. What is a watch? It's a machine that is intended to trace and display the time. This banal piece of common sense leads to another, which has important consequences despite its obviousness. The notion of function or teleology implies the possibility of succeeding *more or less well*. The chair may be more or less comfortable; the watch will be more or less accurate. Every teleological ascription involves a normative or evaluative component. This normative or evaluative element has already come up as a crucial feature of rationality. It applies not only to artifacts but also to intentional acts. An intentional action can be more or less successful in achieving its aims. As a type of action, it is defined not so much by what it *actually* achieves (or even what it is *likely* to achieve) as by what it was *intended* to achieve. The distance between the intention and the result is a measure of its success in the evaluative or normative dimension.

2.2 Teleology in Biology

The compelling nature of the analogy between artifacts and living things is due primarily to this normative dimension. But it is also driven by the apparent improbability of the thought that a given adaptation could have been the effect of mere chance. In a pre-Darwinian context, the apparent improbability of complex adapted structures lends compelling support to the hypothesis of an intelligent creator. As is well-known, the eighteenth-century English apologist William Paley argued that just as we could not avoid appealing to the hypothesis of a watchmaker to explain the existence of a watch found on the beach, so we cannot avoid inferring to the mind of an intelligent creator to explain the creatures encountered in nature, since such creatures are far more complex than a watch. Yet even before Darwin put forward a satisfying alternative hypothesis, this idea met with a fatal objection. If we are to see in the universe the work of an artificer, we must trust common sense to assess the degree of success attained by his creations. And if common sense suffices to perceive that a specific device is the work of an intelligent intentional act of creation, it is also surely competent to notice all the defects that mar that creation. Theologians and philosophers generally refuse to take such notice, on the pretext that human intelligence is too puny to assess the designs of the Almighty. But if that is the case, then there can be no basis for the attribution of perfection or benevolent intelligence to a creator in the first place.[11]

The fact is that despite all the marvels on display in the world of living things, an engineer might find a number of serious flaws. Why, for example, do we not blink in alternate eyes? Blinking with both results in our being effectively blind for about five percent of our waking life.[12] From the point of view of natural selection, working with whatever is at hand under constantly changing constraints, there is nothing surprising about such "imperfections." But the observation reminds us of a question that must be answered by anyone who talks about biological teleology in a naturalistic context. The gauntlet thrown down is this: How, without any reference to any deliberate planning or thought, are we to account for the meaning and notions of *function*?[13] The exclusion of thought precludes any role for deliberate intention in the setting up of a function. But it also signals that we cannot simply deal with the problem by reinterpreting talk of teleology as metaphorical. To take up this challenge, the notions of teleology and function must be explicated without residue in terms of ordinary efficient causality.

The concept of a *selectionist system*, of which evolution by natural selection is the most obvious example, picks up the gauntlet. In the next chapter, I shall compare some typical products of natural selection with some of the strategies used by intelligent agents in devising intentional actions. The comparison will bring out some unexpected points in common.

First, though, it will be helpful to pass in review a time-honored attempt to ground a compromise solution in the idea that goal-directedness derives directly from the organization immanent in living matter. This was Aristotle's solution, and it is still worth looking at, if only to cast the modern view into relief by way of contrast.

Aristotle thought it entirely *natural* to regard the living world as spontaneously organized as if it had been designed with certain aims in mind. On this view, as we have seen, an organism's characteristic activity constitutes the actualization of its potentiality. Here is a way to make sense of this notion of potentiality: among all those events that have never taken place, and among all the characteristics that do not now actually exist, some belong to a privileged class of *nonexistents that are supposed to come into existence*. These privileged nonexistent items—like the oak that the acorn is supposed to become—constitute the *potentiality* of the organism in question. Given an assumption that species are essentially fixed in their nature, which Aristotle is generally regarded as having taken for granted,[14] the study of natural phenomena provides empirical evidence about which events, among all those that have not taken place, can be expected to come into existence "always or for the most part." These will be the events that constitute the actualization of the organism's potentiality.

In the practice of biological sciences, this method is still largely in use. Studying the details of what actually takes place is essential to the discovery of functions. We must assume that things do in fact happen "normally" much of the time. In addition, we can sometimes observe the causal role played by some specific mechanism or component of a biological system in the accomplishment of the system's manifest role (Cummins 1975; Amundson and Lauder

1994). But the philosophical ground on which this method rests has crumbled with the discovery of evolution.[15] If species have arisen from a long dynamic process that has led from unicellular organisms to the elaborate forms of life familiar to us now, we are forced to infer that what happens "always or for the most part" is not a reliable guide to what is *supposed* to happen. Human beings are the outcome of a huge series of natural experiments, most of which were subjected to merciless destruction. At every stage of natural selection that brought us closer to *Homo sapiens*, those of our ancestors who bore the innovative genome were what Aristotle would have regarded as freaks of nature.

It might be objected that if all those accidental steps in the course that led to human evolution had impeded the proper functioning of the organisms affected, we should not be here to witness it. It is indeed by astronomically rare good luck that not a single one of my own ancestors chanced to be infertile. (Doubtless some of my readers' ancestors were equally lucky.) But that is not to license the inference that each of my ancestors' every characteristic contributed to their fitness: to think otherwise would indeed be to fall into the sort of naive adaptationism mocked by Voltaire in the Leibnizian Doctor Pangloss. Among the properties and effects of any given organ, many if not most neither require nor admit of teleological explanation. The laws of physics and geometry suffice to explain, for example, why no insect is as large as an elephant—since the exoskeleton grows proportionately to the square, while body weight increases by the cube (Vogel 1999). And the laws of chance are enough to explain why the vital statistics of any given population arrange themselves so neatly on a bell curve. Such explanations appeal only to simple efficient causes, mathematical necessity, and chance. They have no place for goals or teleology.

These banal observations should hardly be necessary, but they will be highly pertinent when we attempt to gauge the place of adaptation in the mental capacities of rational beings. Still, the appearance of teleology in biology remains compelling. It needs to be explained.

2.3 The Aetiological Theory of Function

The dominant, if not undisputed, account of teleology in biology is provided by the *aetiological theory of objective proper function*, where *aetiological* means "causal" and refers to antecedent causes, not future end states. The basic idea is that the attribution of a function to certain effects rather than others can be justified without reference to any external goal or design. At most, one needs only the entirely general and quasi-tautological premise that the "goal"—in a sort of vestigial sense—of life is to perpetuate itself. The aetiological theory makes use of two insights that I have already mentioned. First, functions and more generally teleological properties *explain* the presence of the organ or phenomenon to which they are ascribed, not in terms of their cause, but in terms of their effects. Second, such properties are akin to Aristotelian potentialities, insofar as these allow us to privilege certain nonexistent effects over others as being *supposed* to take place. But both these ideas take on an entirely new look in the context of an analysis that refers only to efficient causation, in which effects never precede their causes.

Let me turn first to the intuition that the existence of an organ is explained by its function. In order to distance this from the erroneous conception of teleology as backward causation, we must stipulate that the effects to which the explanation refers are not the future effects of a particular organ, but those average effects that resulted, in the population as a whole over the span of phylogenetic history, from the workings of that *kind* of organ. The explanatory force comes from the hypothesis that those effects contributed to the capacity of the present organism's ancestors to transmit organs of that same kind to their progeny. It is plausible to assume, for example, that the function of the heart is the circulation of the blood and not the rhythmic sounds it produces. The reason is that the former, but not the latter, comes into the explanation of the fact that current-day mammals are equipped with a heart. This explanation carries with it a normative force, licensing the claim that if this heart loses its capacity to pump blood, it will no longer be doing what it is *supposed* to be doing. If instead it ceases to make any noise, that fact will doubtless be deemed curious but will not count as a malfunction.

The most sophisticated elaboration of the aetiological theory of function is due to Ruth Millikan (1984, 1993). The following formula (P) is a somewhat simplified definition of the notion of "direct proper function"[16]:

(P) An existing element X has the direct proper function F if and only if:

1. X results from the reproduction of an antecedent element, *ancestral X*;
2. *Ancestral X* effected F in the past, in virtue of properties reproduced in X
3. X exists now because *ancestral X* effected F.

Note that the "because" in (3) is meant in the straighforward sense of a historical cause. The idea is that specific effects produced by *ancestral X* made a difference to the chances of X turning up now. This formula accounts for the main intuitions behind the common notion of function, as used in the context of artifacts and of biology: in particular, it allows us to understand the presence of an organ in terms of what it does and to make the distinction between functioning and malfunctioning, as well as that between functions and accidental effects. More generally, it connects with the broader notion described in the next section.

The Generality of the Selectionist Schema

The aetiological conception provides a basic schema in terms of which a number of "selection-based" systems can be described. Such systems are characterized by three essential features:

1. There is a set of variable elements, the diversity of which can be further increased in some fashion that characteristically is likely to involve chance.
2. There is some sort of selection among the elements mentioned in (1).
3. There is a way in which elements are *preserved* or *transmitted*.

So formulated, the selectionist schema applies equally well to learning as "operant conditioning" described by B. F. Skinner (1953). It also applies to certain "epidemiological theories" of culture, including "memetics," or the theory of memes. Such theories view cultural phenomena of all sorts—concepts, ideas, fashions, inventions, or linguistic elements such as the very word *meme*—solely in terms of their capacity to propagate themselves by becoming "parasites" of human minds. Memetics rather neatly explains its own success as being independent of its own correctness. For the success of a meme is dependent only on its capacity to replicate, and independent of its justification. The parallel among the three domains of application of the selectionist schema is summarized in table 2.1.

Where the characteristic effect of gene G, behavior B, or meme M is itself a partial cause of perpetuation of G, B, or M, it is natural to speak of those characteristic effects as functions of G, B, or M. But the schema does not fit all three cases in exactly the same way. To begin with, the schema says nothing about the genotype-phenotype distinction, which has no analogue in the cases of learning and memes. *Phenotype* designates the individual characteristics of organisms, resulting from developmental processes, controlled by

:: Table 2.1. Three Domains of Application of the Selectionist Schema

Essential Feature	Natural Selection	Operant Conditioning	Memetics
(i) Unit source of variation	Genes (G)	Random operant behavior (B)	Meme (cultural unit) (M)
(ii) Mode of selection	Elimination of less fit	Increased probability of operant behavior	Ease of assimilation by human minds
(iii) Mode of preservation	Heredity of transmission	Memory	Cultural transmission

the genotype in collaboration with other factors, notably environmental ones. Indistinguishable phenotypes can issue from different genotypes, and identical genotypes can give rise to different phenotypes. Only phenotypes are visible to natural selection. As I shall argue in chapter 4, however, it is genes, not phenotypes, that are the main beneficiaries of selection. For only genes are actually perpetuated through the generations.

The cases of learning and memes give rise to a different kind of problem. A moment's consideration is enough to notice that the sense of function defined in formula (P), while preserving some of our main intuitions, does not fully match our usual understanding of the word. Ideas are memes. But the successful transmission of memes is not governed by the criteria of logic, rationality, or cogency of evidence that are assumed to be relevant to ideas in philosophy. As we have all too often occasion to deplore, the most popular ideas, images, ideologies, or beliefs are not necessarily those that their intellectual value, aesthetic merit, or social usefulness would recommend.[17]

In a moment, I shall consider some standard objections faced by the aetiological theory. But first, it will be useful to contrast the selectionist schema with the closely related but distinct notion of feedback.

The Cybernetic Model

Watt's governor is an example of a purely dynamic cybernetic device based on *negative feedback*. As I showed earlier, the same task could be accomplished using a feedback mechanism implemented by SPEEDWATCH, a digital computer. The behavior of the machine would be measured at specified brief intervals and compared to a threshold value set in advance. As the value in question approaches the threshold, SPEEDWATCH would cut down the supply of heat to the boiler and thereby lower the pressure. This is negative feedback. It has much in common with the selectionist schema: it involves a source of behavior that can be more or less random at the outset; the feedback effects a sort of selection, in the sense that it results in the damping of the motion as it nears its threshold value. But there are also important differences:

First, there is no single step that is analogous to transmission or conservation. Feedback acts dynamically on some phase of the process, not on the cycle as a whole. Nor does it act, like natural selection, on the basis of a statistical average based on a large number of individual events.

Second, the cybernetic device involves no source of new variation. It operates on the basis of a fixed repertoire of behaviors that depend on the fixed mechanical properties of the machine.

By contrast, the field of possibilities in biology, despite the constraints to which all biological processes are subject, is importantly unpredictable if not radically indeterministic.

This is related to the third and most important difference: in the cybernetic machine, there is a goal fixed in advance. In biology, there is no concrete goal, other than what I have called the "vestigial goal" of repetition or propagation.

This is why it has proved so tempting to think of life and behavior as transcending the constraints of natural law, while being in some sense polar opposites of merely random events. Typical of this attitude is that of the French philosopher Henri Bergson who, in a book entitled *Creative Evolution*, declared that life appears to us as "an uninterrupted outpouring of new forms" (Bergson 1944). A century of biological advances later, it is now easier to perceive the power of the selectionist schema—though it even now meets resistance. For although that model is childishly simple in principle, it explains how such an "outpouring" can be generated by nothing more than the perfectly tuned collaboration of chance and necessity.

::

2.4 Problems for the Aetiological Model

I have claimed that the aetiological conception of objective function accounts for the most important of our intuitions about teleology. But some prefer to think that the ascription of functions to parts of living things is no more than a metaphor. Their reason is that talk of norms or values presupposes conscious beings that prescribe them or subscribe to them. This presupposition obviously doesn't hold in the case of natural phenomena. Still, between the aetiological criterion for the presence of a function and common examples of functions, the fit is pretty close.

Close, but not perfect. The aetiological conception is essentially oriented toward the past, and that makes trouble for it in several ways. What has served an organism or its fecundity in the past is not necessarily useful in the present. Hereditary traits, like acquired habits, may fulfill all the conditions set out in the aetiological conception in terms of the history that has put them or kept them in place, and yet may now be useless or harmful. This is true of anatomical vestiges such as the human appendix or the legs of certain snakes and also of habits of behavior acquired in one environment that prove useless in another, whether on the scale of an individual life or on that of phylogeny. When food was scarce and calories at a premium, a craving for fat and sugar could aid your chances of survival; nowadays, it is more likely to lead to obesity from an excessive consumption of cheeseburgers and fries.

So is the aetiological conception's backward-looking feature a serious problem? Actually there are two objections tangled together here. Once distinguished, one can be seen to be a pseudo-problem. The other is a real problem but is easily solved with a minor amendment.

The false problem derives from the fact that a capacity that is generally useful to members of a given species can easily give rise to harmful consequences to some unlucky individual. Such events are properly regarded as *accidents*, however, and are therefore incapable of shaking the essentially statistical basis of a function attribution.

The mistake of thinking otherwise is neatly illustrated by a well-known alleged counterexample first thought up by Christopher Boorse (1976). Imagine that in some lab a technician has rigged up an apparatus that involves a gas tap at the end of a tube. A leak in the tube has allowed the gas to escape, and the technician is asphyxiated. Furthermore he is prevented from fixing the leak precisely by that fact: the inability of the technician to fix the leak results from the very condition that it has caused in the immediate past. Since this appears to conform to the aetiological schema, should we then declare that the asphyxiation of the technician is the *function* of the leaky tube?

No. Boorse's mistake is to invoke the causal structure of a particular, one-time-only sequence of events. In a cogent counter-example, the technician's asphyxia would have to perpetuate not just this particular leak, but a whole lineage of leaky tubes, each generation of which would in turn have to cause asphyxia resulting in some technician's inability to plug the leak.

So much for the pseudo-problem. Here is the real one. Common sense, which requires that we make a distinction between functions and other effects, also demands that functions be good for something. By definition, vestigial organs aren't good for anything. So they shouldn't be classed with functional ones. Yet they were useful in the past, presumably contributing to the viability of our ancestors in some given environment in which they had been selected. They exist now in virtue of their past effects: on the aetiological criterion, then, the vestigial legs of snakes still have the function of locomotion, though they no longer serve that function. Do they then *malfunction*? That's not right either: for a malfunction is a failure of something useful to do what it is supposed to, and we've agreed that the snake's legs are useless. Call this the *backward-looking problem*.

Actually the solution to this is simple, if somewhat ad hoc. We need only interpret the third clause in principle in (P) to specify that the effect produced by the trait or organ *ancestral X* must have been recent enough to influence its fitness in the recent past.[18]

The other side of the backward-looking problem is a more interesting one, concerning *future* functionality. What has no function now may acquire one.[19] Even though we won't know this until much later, when the effects of a given capacity will have increased the fitness of the lineage in whose members the capacity is inherited, should we not say that the function is there from the moment it begins to be useful?

This view has given rise to *propensionist* interpretations of functions, on which a function is forward-looking, not backward-looking. Propensionists defend the view that what matters is not the history of the organ but the future consequences of its present capacities.[20] A common argument for this view is that what should be at issue are the present and future causal powers to which we ascribe the status of a function. But why should this be so? The current capacities of an organ and the causal laws in accordance with which they work suffice for the causal explanation of the effects in question. The causal explanation is no less adequate whether or not we add that the effects in question should be described as *functions*. What the notion of function contributes is a factor that goes beyond the strictly causal sequence, namely, the normative or evaluative fact that such events are *supposed* to occur, whether or not they do in fact occur. In other words, the work done by the notion of function consists in recruiting the normative or evaluative dimension. When the heart ceases to pump the blood, we don't say that its functions have changed, but that it has ceased to function. Conversely, a process occurring for the first time that happens to prove useful does not yet count as a proper function. In either case, the functional explanation concerns not the actual causal sequence, but the question of whether it was supposed to take place.

2.5 Teleology against Value

The selectionist schema allows for the ascription of teleology as an objective property of the biological world. As soon as it is applied to social and psychological realities, however, the schema confronts what I shall refer to as *the multiplicity of values*. Fitness constitutes the universal economic criterion of biological value, which sorts effects into those that count as functions and those that don't. I have argued that it is convenient to refer to this as a sort of "vestigial teleology," amounting to no more than the universal and trivial "goal" that underlies all other goals and purposes in the living world, namely, the simple propagation of "replicators." This term does not necessarily refer only to genes in the narrow sense. It denotes whatever is capable of being faithfully copied over a long period of time. But the criterion that picks out vestigial teleology does not necessarily help to discern the multiple and sometimes conflicting values espoused by individual humans.

It is important not to confuse the problem of the multiplicity of values with an objection that is often adduced against the use of economic language in biology. Concepts drawn from economics and game theory are often said to be mere social constructions, and their transfer to biology is suspected of reflecting nothing beyond ideological preconceptions. The complex system of economic exchange modeled by economics and game theory, the objection goes, is nothing but the resultant of the individual and collective choices of humans; and choice can have no more than a metaphorical sense in the biological sphere. Surely, then, the application of economic terms to biology is nothing but the projection of capitalist ideology nourishing an unhealthy kinship with the racist and elitist themes of social Darwinism.[21]

Actually the exact opposite is true. To put the point provocatively: *economics literally applies only to biology*. It does not apply to behavior unless human choices are idealized out of recognition. Economic concepts, such as those of "interest," "benefit," and "loss," get a grip on human behavior only on condition that we take for granted a number of auxiliary psychological assumptions. We need to assume, for example, that people are motivated by self-interest and are disposed to work out the optimal means to the satisfaction of their desires. As to the first assumption, the perusal of any daily newspaper will immediately show how naïve it is to think people follow their interests rather than their passions. (This presumes that we can make the distinction, which is also dubious; but its falsehood will not lend accuracy to economic forecasts.) As for the presumption of rational calculation, we shall soon see that there is much evidence against it. In contrast, when economic concepts are applied to biology, none of those questions arise. The replication of the gene automatically provides a correlative for the notion of interest, and there no issue of whether some deeper or more authentic level of interests might be pried apart from that one.

If some psychological or social trait T is perpetuated in virtue of certain of its causal properties, it is a mere tautology to say that those properties have contributed to the propagation or maintenance of T. That simple fact is what fits economic notions for use in the description of biological phenomena (Maynard Smith 1984). But it is also what gives rise to the problem of the multiplicity of values. For it is all too obvious that the "success" of T may not stop it from violating aesthetic, moral, or other values that we might be concerned with. All too many examples could be given of regrettable aspects of our genetic heritage: predispositions to selfishness, violence, ethnocentrism, or sexism can all plausibly be assumed to be or at least to have once been functional from the strictly biological point of view.

The conflict between values is also illustrated by a proverb dear to economists: *bad money drives out good*. This aphorism would be a mere contradiction in terms if memetic propagation was the only source of value. For what then could be the values in the name of which the successful propagation of the so-called bad money is deplored? Whatever the grounds of our disapproval of "bad money," they must necessarily be distinct from the success of its proliferation.

In the next chapter, I shall further explore the hypothesis that thought generates new possibilities of value, ones that require no justification in terms of biological teleology. By their very nature, such values can conflict, not only with the vestigial value of replicator propagation, but also with one another. They can even be incommensurable, and so give rise to insoluble conflicts.

Without which, after all, life would be the less interesting.

Chapter 3 ⁝⁝ What's the Good of Thinking?

*Care only, in yourself, for what you feel is
nowhere to be found but in you, and make
of yourself, patiently or impatiently—Ah,
the most irreplaceable of beings!*

—André Gide

Neither my unicellular forebears nor the zygote I once was can plausibly be thought to have been endowed with the capacity for thought. It follows that there must have been a first thought, no less in the evolution of the species than in the development of the individual. Before that point, behavior was typically controlled by tropisms or other such mechanisms devised by natural selection. The condition attained by those who have crossed that threshold is often assumed to be a superior one in which behavior is guided by reflection and intention, in virtue of which we deem it fit to bestow upon ourselves the title of "rational animal." As we have already seen, the word *rational* in that phrase is not to be taken in the normative but in the categorial sense. What it tells us about any creature so labeled is that such a creature is *capable of irrationality*, a distinction not bestowed on plants or on less complicated animals. As for members of our own species, before the critical transition to the "age of reason," we can apply normative judgments to their behavior only in a very anemic sense.

But what exactly is meant by "reason" here? In chapter 1, I claimed that a rational method must, in virtue of the very meaning of the phrase, be such as to increase our *chances* of success, whether the goal aimed at be an epistemic or a practical one. What success will amount to, however, and how those chances are computed, will vary from one domain to another, according to the norms of rationality appropriate in any given case. In chapter 2, I advocated a restricted, naturalistic notion of teleology, along the lines of the aetiological analysis of that notion. The argument for endorsing such a notion is that it allows us to rehabilitate a sufficiently robust sense of objective function. That will enable us to give meaning to questions about the utility or value of functions, but it will not suffice to answer such questions. To do that requires a broader understanding of the sources of value.

Such is the basic framework in terms of which I now return to the question asked right at the outset of this essay: What is the good of thinking? The first answer to come to mind is likely to be one inspired by Karl Popper's felicitous remark that "rational method consists in letting our hypotheses die in our stead" (Popper 1972, 248).

This reply has much to be said for it. Thinking allows a kind of experimentation at the level of the individual, without sacrificing that individual and without having to wait for the verdict of selection on the phylogenetic scale. But there are less obvious consequences of this idea to which it is worth drawing attention. Here's what I shall argue: rational thinking actually results in humans confronting a radically novel range of possible *values*. This results from the distinctiveness of individuals and from their capacity for a kind of metacognition that is available only to beings possessed of some sort of language. These fresh possibilities arise in part from the emergence of a multitude of individual ends, articulated through the digital, abstracting tools of representation that language affords. These tools allow the individual to offer a more or less anarchic resistance to the impersonal destiny embodied in the vestigial teleology of natural selection. In a nutshell, then, the argument of the present chapter is this: *thinking, in the fullest sense, generates the radically novel possibility of a plurality of values.*

3.1 Subhuman Rationality

We are never quite done assimilating the consequences of our evolutionary origins. We tend to assume, for example, that there must be a vast chasm separating rational thought from the essentially "mechanical" processes of natural selection. Thought is the quintessence of our superiority, as beings endowed with minds, over the inanimate world. It is the very emblem of the human difference. And yet, if we set out to characterize that vast difference, one might be struck rather by a variety of unexpected parallels.

Look first at the most superficial level of appearances. Abstracting from the issue of temporal scale, the pattern of change among products of technology and the pattern of change in the products of natural selection are remarkably similar. We see everywhere a series of "tinkerings" (Jacob 1982) that gradually transform a type of object until the accumulation of scarcely perceptible changes amount to unrecognizable novelty. Compare, for example, a set of pictures of model changes in a brand of automobiles over the last hundred years with the transformation from *Eohippus* to *Equus Cavallus* over a million years or so, such as might be displayed in any museum of natural history. Both show tiny intermediate steps between any two stages, but a huge difference between the models figuring at either end of the process. As Tim Lewens has remarked,

> In both cases this set of available solutions is constrained; developmental factors dictate that no gun-toting zebra variant will arise to answer the selection pressure of evading

> predators, and equally facts about an individual's cultural environment will dictate that only a handful of possible solutions will be considered to a design problem. Although the extent of constraint may be greater in the organismic case, it is surely true that an artificer's cultural heritage will predispose them to ignore some solutions and focus on others. (Lewens 2002, 8–9)

That point of resemblance may strike you as superficial. But there are deeper analogies. As we saw in the last chapter, the power of natural selection is not only manifested on the phylogenetic scale. Essentially the same explanatory schema applies to learning by operant conditioning and to the replication of memes. It can also be seen to apply to connectionist models in artificial intelligence (Edelman 1987). And in embryogenesis, the elimination of cells by apoptosis or "orderly cell suicide" plays an essential role in the elaboration of developing structures, including that of the brain.[1] "Genetic algorithms" provide another increasingly important illustration of the power of the selectionist schema: more or less random lines of code (forming an original source of diversity) are subjected to a process of selection, then randomly mutated or recombined, and subjected to selection again; in this way, the algorithm produces functioning programs that achieve some desired result with remarkable efficiency despite the absence of programming.[2]

An interesting illustration is provided by the collective behavior of an anthill. Anthills are remarkable for the wealth of capacities they have developed on the basis of essentially selectionist mechanisms. But they also illustrate a significant deficit of the power of such mechanisms in comparison with those of deliberate thought.

If you place a source of food at some distance from an anthill, a busy trail of ants will soon form along the shortest available path from the anthill to the food, going around any obstacles placed between them. Individual ants are presumably quite incapable of even the most rudimentary practical reasoning. So how do they solve the problem and find the shortest path? They behave, it seems, as if they were implementing the following algorithm:

1. Follow the path with the greatest concentration of pheromone, if any.
2. Otherwise walk around at random for a while, then return home.
3. If food is found, return home, secreting a pheromone to mark your path.

Together, these three rules suffice for a collective solution to the problem. The reason is easy to see. At the beginning, all ants explore the neighboring space at random. Some will be bound to hit on the food source, and some of those will have happened on the shortest path to it and return by the same path. Assuming that all ants travel at roughly the same speed, those that take the shorter path will get there first, and they will return first along path. That path will therefore have the strongest scent for others to follow. This will set up a positive feedback mechanism, whereby that strongest of pheromone tracks will be followed by the largest number of ants, which in turn will enhance the pheromone track. At the level of the community, we can think of this as setting up a kind of selective reinforcement, analogous to that which operates at the individual level in the case of conditioning. The establishment of a "best path" conforms to the basic schema set out in the previous chapter (see table 2.1, page 47), since it involves a *source of variation* (2), a principle of *preservation or transmission* (3), and a kind of *selection* (1).

The resulting mechanism is marvelously effective, but it will not find an optimal solution when the anthill is faced with two equally short paths. The intelligent solution in that case, as any rational agent equipped with a topographical chart would determine, is to distribute individuals between the two paths, in order to avoid one path becoming overcrowded. But the ants will never discover this solution. By chance, one of the two paths will always be more traveled at the outset, and that path will attract increasingly many ants by virtue of its stronger chemical message. The ants will be caught in a traffic jam while other equally desirable paths remain virtually deserted.

This drawback of the ant algorithm brings out a second, often mentioned, difference between the possibilities of thinking and those afforded by natural selection.

Whoever has walked in the mountains has had the experience known in German by the lovely name *Falschspitzfindigkeit*: unless you can look at a topographical map, you will mistake nearby hills for the summit you are aiming for. This idea has inspired the image of a *fitness landscape*, in which like hikers in hilly country organisms continually change their position in the hope of reaching greater heights of fitness. The highest peaks represent the highest degrees of fitness to which the organism can aspire by dint of changes in its phenotype. The process of thinking enables one to survey available alternatives ahead of time on a real or figurative map. Thanks to this, we can plan ahead and take account, when appropriate, of the need to retreat in order to secure a better starting point for further progress.

But natural selection has no maps. It can only follow, myopically, the line of rising altitude; necessarily, then, it will get stuck on the foothills, or "local maxima." The only way it can avoid getting stuck in a local maximum is to go in for occasional random wobbles instead of aiming for higher ground at every step of the way. That allows it to spring out of the path leading to the nearest hilltop and thereby, with luck, to reach higher peaks available only from some lower point in a nearby valley.[3]

In practice, the ants, too, find ways of wobbling at random. Or rather, the anthill does: for remember that it is the anthill as a whole, not any individual ant, that acts in ways that are putatively rational. In order to mimic intelligent behavior in the sorts of situations just mentioned, the anthill can exempt some individual ants from its normal rule. If some ants persist in foraging at random and ignoring rule (3), they may not do well individually, but they will keep alive the prospect of discovering alternative new paths for the anthill as a whole, particularly once the present source of food runs out. By ignoring the rules, the deviant ants will ensure that there is enough wobble in the collective behavior to provide for adaptation to new situations when they arise.

What provides that kind of "wobble" at the level of the genome? The likely answer is intriguing. Some genes, homeobox or Hox genes, are known to have been conserved in astonishingly constant form for so long that they continue to be present in organisms belonging to vastly distant phyla. Genes of this kind, for example, regulate the segmentation of insect bodies in accordance with their basic body plans. Small alterations of these genes can cause profound changes on the phylogenetic level as well as in the development of the individual organism. Some mutations, for example, cause whole organs to be displaced or replaced. A well-known example is the complex Antennopedia, in the fruit fly *Drosophila melanogaster*, studied by Walter Gehring and others. Changes in the complex cause the fly's antennae to be replaced with an additional pair of legs. To the surprise of biologists, the same regulatory genes, made up of essentially the same sequences of DNA, have been identified in vertebrates (Carroll, Grenier, and Weatherbee 2001; Gehring 1999).

The discovery of Hox genes casts a new light on homology and diversity. We can now glimpse how it is possible for the overall body plan of an organism to be modified in a coherent way on the basis of a small change at the level of genes. A small mutation in the sequence of DNA bases can lead to a leap at the level of phenotypes.

This is one source of the crucial "wobble" that allows natural selection to transcend its essential myopia.

3.2 Deliberation and Selection: Three Cases of Supposed Divergence

Elliott Sober has identified three types of cases in which the solution found by rational deliberation diverges from the probable equilibrium arrived at by selection.[4] As we shall see, his argument doesn't quite establish his conclusion. Of the three cases in question, the first doesn't really manifest the divergence he claims to detect. The second is described in excessively abstract terms, and the conditions under which it might occur are unlikely to be realized. If they were, moreover, the resulting cases would tend to show not that deliberative thought does better but, on the contrary, that natural selection can arrive at solutions that systematically elude deliberation. The third type of case alone is a genuine illustration of the superiority of rational thought in solving certain problems, and its advantage here seems to be entirely due to the possibility afforded by linguistic representation of formulating counterfactual hypotheses.

All three of Sober's cases are based on variants of the *prisoner's dilemma*. This term refers to a vast range of interpersonal situations that share the following startling feature: *if all participants rationally pursues their own best interests, the outcome for all is worse than it might have been had each chosen otherwise.*

We can see how this works in the example that gives the puzzle its name. Imagine that two prisoners—never mind their guilt or their crime—interrogated separately, are presented with the choice:

(PD) If you confess and your accomplice does not confess, you will be released and he will have ten years in prison.

If you both confess, you will each have five years. If you both refuse to confess, we will still find a way to lock you up for a year.

The specific figures set out in table 3.1, are of course chosen arbitrarily for the purposes of illustration. All that matters is the differential payoff between the choice to "cooperate" and the choice to "defect." We can see from the table that each player, having no knowledge of what the other player will do, will conclude that it would be better to defect no matter what the other does. The choice of defection is said to *dominate* the choice of cooperation.[5] Both players will reach the same conclusion, so both will end up in the situation presented in the bottom right hand frame of table 3.1: they both will get five years. For each one of them, that's only the third best (or second worst) outcome.

Sober's argument rests on the parallel between biological fitness and the utility of players in a PD game. In terms of fitness, cooperation and defection represent altruistic or egoistic behavior respectively. Sober maintains that in the biological case, organisms competing in a PD-like situation will not necessarily land themselves in the same frame of the table as two players applying the principle of dominance. The reason is this. The consequences of

∷ Table 3.1. The Prisoner's Dilemma (Ordered Pairs in Parentheses Give Expected Utility for Player A, Expected Utility for Player B)

	B cooperates	B defects
A cooperates	(−1, −1)	(−10, 0)
A defects	(0, −10)	(−5, −5)

a strategy depend on the strategy of the other players. And agents may be considerably more likely to find themselves dealing with others similar to themselves. Thus, altruists may find themselves regularly amongst altruists, egoists amongst egoists. Since we are considering only the objective behavior of the agents, and not any sort of rational calculation, two altruists could accidentally form a "society" that places them in the first frame of the table, where both end up better off, with the second best among possible outcomes.

This line of thought is quite correct, but it falls short of proving Sober's claim that the outcomes produced by deliberating agents and by natural selection will diverge. Agents who deliberate are just as capable of taking into account a high probability that their adversary will cooperate, and so can reach the same mutually beneficial conclusion in favorable cases. So the moral is not that there is an intrinsic difference between the capacities of selection and deliberation, but rather that other factors, whether these be features of the selective environment or additional premises in deliberation, can intervene to soften the dilemma. It remains true that each player, in ignorance of the probable intentions of the other, can make no better rational choice than to pick the dominant option. A player can decide to cooperate, however, in the light of additional information of a sort that involves the kind of second-guessing or "metacognition" that requires a capacity for explicit thought.

In the case of nonhuman organisms, on the contrary, the objective situation in which two altruistic agents could form an alliance does not arise because of the information at their disposal. It simply results from some assortment of similar organisms into separate groups. So there is indeed a difference between deliberating and non-deliberating agents. But it is not quite the difference Sober describes, since it is due to external factors that may influence either outcome, rather than to any difference in the logic of the two implementations—with or without intention—of strategies of cooperation or defection in the prisoner's dilemma.

To prepare the ground for the second of Sober's cases, consider the paradox of the "surprise examination." A teacher announces that there will be an exam sometime this week and adds that the students will have no way of knowing when it will take place until that very day. A cunning student reasons that no exam meeting the teacher's specifications can ever take place. It cannot take place on Friday, the last school day in the week. For on Thursday, students will be in a position to deduce that, since there is only one day left in the week, the test can only take place on Friday. That would give the lie to the teacher's claim that the students will not know the time of the test before the day itself. Since Friday is out, by the same argument the test cannot take place on Thursday, or Wednesday, and so on to the first day of the week. The cunning student's "regressive induction" seems unimpeachable. She is all the more surprised on Wednesday morning, when the exam does indeed take place, fulfilling all of the promised conditions.

This situation derives its paradoxical character from the induction that gives the student a false sense of security. If none of the students had come up with that argument, they might still have been surprised, but there would be no paradox. And it is precisely because of the paradox that the teacher's announcement rules out the possibility of making a rational prediction, at the level of metacognition or reflexive deliberation.

A similar "regressive induction" is involved in Sober's second kind of case. Once again metacognition plays a critical role. This case deals with an iterated PD of fixed length. (In an iterated PD, the same players play the game a number of times.) Both players know that in an iterated PD no strategy dominates the "tit-for-tat" strategy (Axelrod 1984). By cooperating in the first round and then matching the other's choice in each subsequent round, we can reap the benefits of cooperation indefinitely, but if we are betrayed we will be sure not to play patsy twice. In such a case, then, we can expect that others will pursue a cooperative, or "altruistic," strategy because this will seem as rational to them as it does to us. Come the last round, however, it will be too late for reprisals: so now it will be rational to defect. That sets each player up for a regressive induction similar to that of the surprise test. If it is rational to defect in the last round, each will realize that the other knows this too, and so will rush to defect first. On that basis, assuming the other will defect at the last round, each player will rationally defect in the second to last round. And so on back, to the point at which we persuade ourselves that we would be better off defecting right from the first round.

Now, says Sober, consider a three-round game. There are four strategies corresponding to all possible permutations. The first (T3) plays tit for tat in all three rounds. The polar opposite (T0) is that of the pure egoist who defects in each round. Between these two poles, there are two other possible sequences: (T1) plays tit for tat in the first round and defects in the second, while (T2) plays tit for tat in the first two rounds and defects in the third. The winning strategy against (T3) is (T2); the winner against (T2) is (T1), and (T1) is bested by (T0). But what is the best strategy *overall*, assuming neither player knows anything about the other? Sober makes two claims about this. The first is that no principle of standard "Bayesian" decision theory can determine the best strategy. Decision theory bases itself on "expected utility," or the mathematical expectation of value. The "expected value," V, of an action is equal to the sum of the values of each consequence, v_i, weighted by its probability, p_i:[6]

$$V = \sum_{i=1}^{n} (p_i \bullet v_i)$$

But unless we can estimate p_i, the probability that the other will adopt each of the four possible strategies, V remains indeterminate.

Now Sober's second claim is that if we model the effects of selection in such a sequence, we do indeed get a stable result. (T2) will beat (T3) but will be defeated by (T1). But (T1) will succumb to (T0), which is the only stable strategy (Sober 1998, 412–14).

Here, then, deliberative logic and the "logic" of natural selection really do diverge. Curiously, this divergence is far from vindicating the superior powers of thought. On the contrary, as in the case of the surprise test, we find here that *it is precisely the capacity for reasoning that leads to the agent's paralysis*. Given that the participants can make no rational estimate of the pertinent probabilities, the decision algorithm gets no grip, nor does the principle of dominance. Selection does not suffer from this problem. It does its work and we can expect it to zero in on the stable "solution" (T0) in the majority of cases. (But don't forget that the stable solution is not the best outcome.)

Deliberative thought does, however, come into its own in Sober's third kind of case. This concerns the effect that an altruistic act can have on the future of an individual who performs it, in virtue of the individual's membership in a group that benefits from the act in question. The critical factor in such cases is the capacity to formulate counterfactual judgments on the future effects of different strategies. Natural selection lacks this capacity to calculate the possible consequences of alternative strategies. It will thus be unable to recognize the long-term desirability of a choice that offers present disadvantages but future benefits.

Thus, explicit deliberation allows for a more sophisticated level of rationality, by providing access to information not locally available. That's how the topographical map enables our mountain hiker to avoid *Falschspitzfindigkeit*. Language enlarges the field of possibilities because it makes it possible to envisage counterfactuals.

Here, at last, we have a case in which the advantage seems to be firmly on the side of thinking. However, we will see that, in certain conditions, natural selection, too, has the power to open up new worlds of possibility.

3.3 The Multiplication of Possibilities

But what, in that last sentence, could *possibility* possibly mean?

We need to distinguish between two very different kinds of possibility. One I shall call *general possibility*; the other, *singular possibility*. The first term typically applies to general, atemporal propositions. We can ask if it is possible, in that sense, that a particle should move faster than the speed of light, or if a number could be both a square and a cube. Singular possibility, by contrast, designates what is possible *for* some particular person or thing, at a given place and time (Prior 1968). Some things are possible for one person at a specific time, but not for another person, or not for that person at another time. Tim Molloy, in *On the Waterfront*, might once have been a contender, but by the time he got into that taxi, that was no longer possible.[7] Some things are no longer possible for me, which once were possible. At one time I might, like Achilles, have met with an early and heroic death: but now it is too late.

Every singular possibility corresponds to a general possibility, but not vice versa. The singular possibility just noted, for example, could be reformulated in general terms: *That Ronnie should meet an early heroic death in 1960* is a possible proposition; *That Ronnie should meet an early heroic death in 2006* is not. A general possibility, by contrast, is equivalent to a singular possibility only if it relates essentially to some particular individual at a specific time. The possibility that unicorns existed does not relate to any individual, existing or extinct. It is a purely general possibility. (Or perhaps it is an impossibility, if, as Kripke once suggested, unicorns are essentially mythical. For it would then follow from the concept of *unicorn*, just as it follows from the concept of *simultaneously instantiating all perfections*, that no entity could instantiate it.)

A widely favored way of conceptualizing possibility in modal logic is based on possible worlds. In those terms, we could say that singular possibility tracks an individual across possible worlds accessible from his original world.[8] Every event in the world of my life is a historical phenomenon, and these events do not simply play themselves out in a domain of fixed possibilities, but rather create new singular possibilities. Over the course of evolution, of history, or of individual life, it is not just facts that come into being but also singular possibilities (Jacob 1982).

The Critical Transitions of Evolution

The idea of changing possibilities is nicely illustrated by Maynard Smith and Szathmáry's eight "critical transitions" in the history of life (1999). Each marks the advent of a new range of possibilities. The first made possible the phenomenon of life itself: this was the formation of autocatalytic systems, molecules capable of mutually catalyzing their duplication. Without mutual catalysis, the duplication of molecules—that is, life itself—was not a concrete singular possibility for any existing entity. Two other transitions among those studied by Maynard Smith and Szathmáry are of particular relevance to my argument because the growth of possibles to which they gave rise takes place on the level of *information*. One is the creation of the genetic code. From that point on, reproduction relied on RNA and DNA molecules, which lead a double life as physical models and depositories of information: Should we think of a gene—that is, a DNA molecule—as a structure that is replicated or as information that is copied and translated?

> In present-day organisms, it is both: a gene plays two roles. It acts as a template in gene replication, so that two identical copies are made of a single model.... [In addition, and] by an exact analogy with a tape recorder or a television set, we can say that the sequence of bases in a gene carries information that specifies the structure of a protein, just as the magnetic pattern on the tape specifies the sound that comes out of the recorder. (Maynard Smith and Szathmáry 1999, 10)

Assuming, then, that we are dealing with an essentially digital code, we can count the number of elements and their possible combinations and make a rough calculation of the order of magnitude of the resulting explosion of possibilities. Taking account only, for now, of the protein-building function of DNA, the code needs to have a "word" for each of the twenty amino acids that are the building blocks of proteins. There are four "letters" in DNA, so a sequence of two will not allow for enough differences, since there are just sixteen ways of arranging two such letters. Three letters will yield sixty-four permutations or possible "words." As proteins are sequences of amino acids, the number of such "words" in a string of DNA will determine the number of theoretically possible proteins that the string can specify. Suppose we were to limit all proteins to fifty components. We could already specify, with a chain of 150 DNA bases, 20^{50}, or more than 10^{65}, proteins. To get a sense of what that means, we might note that this order of magnitude is close to a commonly cited estimate for the total number of individual atoms in the galaxy. Because each string of proteins necessarily contains a multitude of atoms, and because actual protein commonly consists not of fifty, but of many hundreds of amino acid elements, we can readily conclude that the world of possibilities brought into being by the invention of the genetic code is practically unbounded.

A similar combinatorial explosion is made possible in the last transition that Maynard Smith and Szathmáry talk about: the transition to language. The remainder of the present chapter will focus on this transition and on its specific contributions to thought and rationality.

3.4 Language and Intentionality

In the same way that DNA provides the means for specifying an astronomical number of possible proteins, the compositional structure of language makes possible a hyperastronomical number of sentences. It thereby allows for the expression of an immense domain of thoughts. But language does not merely permit their expression. As we shall see, language actually enlarges the domain of possible thoughts.

In a recent book, Joëlle Proust reads Descartes as advancing an argument that marks "the contrast between human abilities [and those of other animals].... His argument consists in essence in contrasting the particularity of animal skills with the universality of human thought" (Proust 2003, 40). I don't know whether she is right about Descartes. But there is a sense in which that formula expresses precisely the opposite of the truth.

Proust may have had in mind the fact that animals react in real time to particular circumstances with which they are confronted and lack the capacity to articulate general maxims governing their behavior. This is no doubt true. But if we look at it in a different light, we can see a sense in which the thoughts of animals—or of machines insofar as thoughts can be ascribed to machines—are *essentially general*.

Here is a bare statement of what I mean. (Bear with me: this is compressed; but the remainder of the present chapter will be devoted to making it clear.) In one sense, generality contrasts with the specific. But in another sense, its opposite is the particular. Only the most highly developed form of intentionality is capable of marking the logical distinction between the two—that is, between a *particular* and a *specific* object. This highest form of intentionality is accessible only to minds capable of using the machinery of language. And the capacity to mark that distinction, in turn, is one of the conditions that makes possible the multiplication of values.

To elucidate this dark doctrine, I must explain and defend two theses: the first is a logical one, concerned with the difference between the particular and the specific; the second addresses the question of how there come to be so many potentially incompatible human values.

To explain the logical thesis, we must first take a step back and look at one very special—and very obvious—feature of our human minds. I refer to the fact that we have thoughts *about* things. What kind of relation is it between my mind and Ulysses when I am thinking about Ulysses? Philosophers call this feature "intentionality." Rivers of ink have flowed in debates about how properly to define it. Some progress was promised by an ingenious way of picking out sentences that ascribe intentional states, proposed in the fifties by Roderick Chisholm (1957). His suggestion was borrowed from the Austrian philosopher Franz Brentano (1838–1917), who was himself inspired by some medieval ideas. Chisholm identified certain grammatical tests supposedly capable of singling out sentences intended to attribute intentionality. By the same token, these tests were supposed to pick out sentences that ascribed intentional states to a subject with a mind.

The two criteria Chisholm proposed were intentional inexistence and generality. The former is based on the following contrast. If a physical object affects another physical object in some way, both must exist; the intentional object of a thought, however, can somehow "affect" that thought without existing. I can think of a unicorn, for example, but in the real world, there are no unicorns. So from the truth of "Joseph believes in Santa Claus," one cannot infer the real existence of Santa Claus. By contrast, the truth of an utterance that attributes a physical relation to him—such as "Joseph ran into Santa Claus"—requires him to exist (unless the reference is understood merely figuratively). This is what is meant by *intentional inexistence,* a phrase that connotes both the idea that Santa Claus exists only "in" the mind and that it *lacks* existence.

As for the criterion of generality, it is based on the fact that causal relations can hold only between events whose constituents are *particular objects*. Thought, by contrast, by dint of its capacity for abstraction, can envisage not just nonexistent objects but also objects *in general*. Hence the idea that intentionality implies a sort of generality.

These two tests seem to yield an ingenious as well as plausible way of picking out beings capable of having mental states. So it is disconcerting to realize that both these tests will classify an automatic dispenser of pop drinks as having mental states.

Understanding why requires us to take account of one more distinction, between two flavors of generality. The difference between them is manifested in their antonyms. One takes *specificity* as its opposite, and the other, *particularity*. A general statement can be more or less specific. The more predicates we attach to a concept, the narrower that concept becomes, which is to say that the set of objects designated by that concept shrinks. We do this whenever we play twenty questions.[9] At the limit of specificity, this set can narrow down to a single member—and just one more specification could bar even this unique last member, leaving nothing at all that fits. But unless a predicate is self-contradictory the size of the set it denotes is always contingent. No general specification is logically sufficient to guarantee that the concept picks out a single particular object. Specificity does nothing to diminish generality in the second sense of that term. When we speak about a particular object, that object is unique as a matter of logical necessity. We would be misusing the notion of particular reference if we thought it allowed equivocation between several objects (or if we thought there was no object referred to at all).[10] This would not simply reflect a lack of precision in the term used to designate the object in question. We can refer to a particular object with a term that specifies almost nothing. In a suitable context, the bare indexical *that* suffices.

Armed with this distinction between two types of generality, we can now define "quasi-intentionality," as distinct from the full intentionality of thought, and we can see how it applies to a pop dispenser.[11] Let us assume that this device is designed to dispense a can of soda whenever a $1 coin is dropped in the slot. We could then say that the machine is "ready" to receive a coin. This notion satisfies the criterion of intentional inexistence: even if all $1 coins had vanished from the universe, the machine would still be ready to receive one. It also satisfies the criterion of generality because the machine is willing to spit out a can in exchange for any coin of a certain kind. Its disposition makes no reference to any particular object. So the pop machine is indeed a quasi-intentional device. If this machine does not achieve full intentionality, then, this is not because machines—or animals—are confined to the particular. Quite the contrary: the problem is that they are stuck in generality. What machines and animals lack is the means to apprehend the particular as distinct from the specific.

In order to exhibit full-fledged intentionality by referring to an individual, one must have mastery of a language sufficiently rich to distinguish a proper name from a common noun, and so to grasp the difference between the recurrence of the same object and the occurrence of two qualitatively indiscernible objects.[12]

In fact, the capacity to make that simple logical distinction is intimately bound up with some of our highest values. Our passions are typically directed at concrete individuals, and we hold dear the idea of every human being as a unique, irreplaceable individual. Thus fully intentional thought underlies our capacity to identify and value individuality as such, and that, in turn, depends on the crucial transition effected by the acquisition of language.

Before exploring this idea a little further, let us dwell a little on the relation of language to some of the other essential capacities of thought.

::
3.5 The Modularity of Mind and the Role of Language

In what follows, I shall borrow from Peter Carruthers (2002, 2003) the hypothesis that one function of language is to *mitigate the isolation of our mental modules*. To make this intelligible, let me start by explaining the relevant notion of "modules," and why we should think they figure in our cognitive capacities. The notion of a module originates with Jerry Fodor, who used it to designate the peripheral operations of cognition, such as those at work in vision, hearing, or touch (Fodor 1983). Such modules are typically innate; they work quickly and unconsciously, and their competence is confined to a specific domain. Most modules, in this narrow sense of the term, perform afferent or sensory functions, but some forms of aphasia and dyslexia show that some motor functions, particularly those involved in phonological or syntactical language production, are also modular. These modules typically involve specialized *transducers*, mechanisms that convert information about sound waves, for example, or activity in the electromagnetic spectrum, into neural impulses. They are therefore relatively autonomous neurologically as well as functionally, which leaves them "informationally encapsulated." Informational encapsulation is in evidence in many optical illusions that persist even when we know them to be illusory. When we look at a straight stick immersed in water, we cannot help seeing it as bent at the surface of the water even if we know quite well that it is not. This encapsulation is not absolute, since we can sometimes be brought to see something differently in the light of information. A sketch that consists of a simple circle with two straight lines sticking out of it could represent an olive impaled on a toothpick, but all it takes is a caption for us to see a Mexican on a bicycle seen from on top. (More grandly, Paul Churchland (1979) has shown that with a little training we can literally come to see that the earth turns.) Unlike perception, belief is generally permeable to information: we will tend to give up a belief in the face of compelling evidence of its falsehood. The power of logical and mathematical reasoning derives from the fact that they are *topic neutral*: the validity of an argument depends on its form, not on its content. Similarly, speech, as such, is universal in scope. We sometimes hear people speak of the ineffable, but the domain of the ineffable is not independently characterized. Different topics may call for different vocabularies, but not for a different faculty of language. Logic, mathematics, and language are not restricted to any domain. Nor do they use specialized transducers.

From these obvious facts, Fodor concluded that knowledge and reasoning generally could not be modular in the strict sense of the term as he defined it. But more recently, philosophers and psychologists have relaxed the definitional criteria for being a module, thus making it plausible to speak of cognitive modules, lacking a specific neurological basis, only relatively encapsulated.

There are both theoretical and empirical reasons for thinking that such modules exist. Let us consider first the empirical evidence.

In the Wason test[13] subjects are confronted with one or the other of the two following problems:

1. Here are four cards with one side hidden. We know that each card has a number on one side and a letter on the other. The four visible sides of the cards show the following: D; F; 3; 7. Which cards do we have to turn over to confirm that the following rule holds true: "If a card has a D on one side, it must have a 3 on the other"?
2. You are hired by the owner of a bar to make sure that anyone consuming alcohol is eighteen or older. Each of the following cards represents a customer at the bar. One side lists her age and the other side lists what she is drinking. The four visible sides of the cards give the following information: beer; lemonade; 25 years old; 16 years old. Which cards must you turn over the make sure that no one is breaking the law?

In experiments that have often been replicated, most subjects give the wrong response to the first problem, while most get the second one right. Yet the two problems have an identical logical structure—[*If p, then q*]—and hence an identical solution. The rule is broken only when *p* is true and *q* is false; so we must turn over the first and last cards but there is no need to turn over any of the others. After devising a number of ingenious variants of the experiment to exclude rival hypotheses, Cosmides and Tooby concluded that we have evolved a specialized module whose function is to *detect cheaters*. In a cheater-detection scenario, this module quickly solves the problem. When faced with other problems the ostensible content of which does not call this function into play, however, we go wrong, even when the problem is cast in precisely the same logical form.[14]

What's the Good of Thinking?

Another kind of empirical evidence for modularity is to be found in the observation of the intellectual development of children, as well as in certain pathological cases. In certain forms of aphasia, for example, functions that we would normally assume are inseparable (reading and writing, for example, or understanding verbs and nouns) are dissociated: one is normal while the other is lost. Furthermore, we could cite the curiously weak correlation between different forms of intelligence.[15]

This mere sketch of the empirical bases for the modularity hypothesis will suffice for my purposes. Turning now to theoretical support for modularity, consider first the matter of computational complexity. Generally speaking, a machine dedicated to a single task is likely to perform that task more efficiently than will a universal computer. Its wiring can be simpler, and it may consider a more limited range of information. Our different organs perform a large number of disparate tasks; in one sense, the brain is the "all-purpose" computer, which directs the circulation of the blood, digestion, the immune system, and countless other tasks all at once. But as the pinpointed effects of brain lesions on different functions makes clear, it is not the case that all these jobs are done by the whole brain. Despite the huge connectivity of the brain, in which every neuron is connected to an average of a thousand others, different parts do perform distinct functions or play specific roles in the performance of complex functions.

The same seems to be true for the different domains of cognition. There appear to be, for example, modules that deal with "natural biology" and "natural physics," on the basis of which we form, from a very young age, systematic expectations based on inductions that greatly outrun our experience of living beings or inanimate objects. There is also evidence for the existence of a specialized system for detecting others' states of mind; a geometrical system of navigation; and several systems charged with the management of social relations, of which cheater-detection is just one (Carruthers 2003). This shouldn't be surprising, in that the "tinkering" process of natural selection is unlikely to have had the leisure to put together in one stroke a universal machine capable of performing all the specialized operations that have come to be required at various times over the course of evolution.

I spoke in chapter 2 of an objective vestigial teleology, anchored in the reproduction of genetic forms. But that apparent teleology reveals no unified plan. For that reason, social scientists have tended to regard the biological level of explanation of human behavior as useless. They are right to stress the immense diversity of conscious and unconscious motives that drive human beings. In the last analysis, however, the capacities that make such abundance possible must be rooted in biology. Over the course of evolution, organisms have elaborated all kinds of strategies in the pursuit of their myriad intermediate goals—ranging from the representation of the three-dimensional space in which we navigate to the recognition of desirable qualities in a sexual partner. We should expect these strategies to prove as disparate in their functioning as our bodily organs. And there is no reason to believe that our cognitive modules will always work smoothly together.

On this point, consider one more suggestive piece of empirical evidence. Certain experiments seem to confirm that our rat cousins are unable to transfer the use of pertinent information from one sensory modality to another. Rats were placed in a rectangular cage in which there was a visible food source in one corner. They were then removed from the cage, disoriented, and returned to the same cage, where the food was now hidden. Different areas in the cage were clearly marked by distinctive smells or by bright or dark colors, which these animals' sensory capacities are well equipped to detect. The researchers found, however, that in their quest for the hidden food, the rats used only geometrical cues, ignoring the cage's other landmarks entirely.[16]

Wherein lies the difference between rats and us? How did we become capable (to the extent that we are) of taking account of disparate information in all sorts of different contexts? This question becomes all the more pressing in light of Plantinga's challenge, raised in chapter 1. You will recall that Plantinga, like the God of Genesis, wants to modularize knowledge and cast doubt on the reliability of the "knowledge of evolution by natural selection" module. If we have access only to modules designed to accomplish some particular specialized task or other, and if these modules were formed under concrete selection pressure, why should we expect any of them to help with entirely new problems of a purely theoretical nature?

The answer is that the information furnished by the diverse specialized modules are linked together by a universal, topic-neutral system of representation. Those modules have been running smoothly for millions of years; the universal system, by contrast, has had only a few thousand years to perfect itself. Its development required a new way of representing information. It seems likely that this new system of information processing evolved in two stages, each of which marked a major leap toward abstraction. First, some hundred thousand years back or so, came the invention of *language*. For the purposes of the present argument, what's crucial about language is that it constitutes a *digital* system of representation. The second stage, which is no more than a few thousand years old, was the refinement of that invention to incorporate a *topic-neutral logic*. That is the tool that enables us to solve abstract problems such as the Wason test and, most importantly, to *show* conclusively that the first intuitive solution most subjects come up with is mistaken. One drawback of these verbal, digital, and topic-neutral reasoning methods, however, is that compared to our specialized modules they are excessively slow. Furthermore, though we are capable of obtaining methodically trustworthy results, we must also acknowledge our frequent tendency to get confused when we engage in explicit reasoning, particularly when dealing with probability. But the mistakes we make in our mathematical and linguistic calculations are not *systematic* mistakes, of the kind our specialized modules sometimes produce.

3.6 The Emergence of Individuality and the Diversification of Values

If one abstracts from time, and from the individual identity of biological organisms, the mechanisms of thought turn out to differ surprisingly little from those of natural selection. The difference becomes crucial when we focus on what a single individual can achieve in the span of a single life. I have argued that individuality has two distinct aspects: one logical, one metaphysical. From the logical point of view, the identification and reidentification of a particular as such requires specific linguistic tools. Metaphysically, once we focus on the individual as the *locus of value,* an indefinitely large range of new possibilities open up for the realm of values. Natural selection deals with statistical effects affecting large numbers, in the context of populations and lineages. By contrast, unique events are what matter to the individual.

The metaphysical point depends on the logical one. For once we grasp the difference between "a thing of that kind" and "this particular individual," it becomes easy to see that this individual might have goals that fail to line up with those "goals" of an organism that derive merely from its status as a member of the biosphere. The way is open for a burst of diversity in what is of value to individual human beings.

To flesh out this bare claim, we need to attend to the notion of intention, not in the semitechnical sense of "intentionality" that has come up in the argument so far, but in the ordinary sense of "intending to act." Intention in this sense is the paradigm of teleology. Any intention can be thought of as setting up a *norm of success*, in terms of which the intentional action will be deemed to attain or to fail of its goal. Some of those norms pertain to external conditions of success, while others have to do with the appropriateness of the intention itself in the light of the agent's overall goals. In turn, the notion of appropriateness involved may depend on some other norm or value, which might be instrumental or intrinsic, moral or aesthetic. Those norms and values can be further refined and differentiated indefinitely into further categories. That diversification of norms and values, derived from the instrumental or intrinsic goals all agents set for themselves, is what makes it possible for values to *conflict*.

But surely, it might be objected, conflicts might arise, between or within individuals, without the need for both sides to lay out their case in plain English. True; but the sort of conflicts I have in mind are not such as might arise between two alternative means to a single goal. Neither can they be understood on the model of two animals fighting over a meal. Those are conflicts of interest, not conflicts of value. In order to make sense of genuine conflicts of value, we need to be in a position to specify clearly what it is for a goal to belong to a particular organism, rather than arising from the general, vestigially "teleological" imperatives to which any organism of that sort is subject. Language, as we have seen, is crucial to our capacity to express the thought that we are reidentifying the same individual. Only by means of language, then, can one confirm that one is referring to some particular object by virtue of one's intention to refer just to that object and to no other.

What I have just said does not mean that a dog can't in practice identify its master or more generally that an animal without language can't identify a particular individual. Vampire bats are a case in point. They live in colonies and feed on the blood of mammals; and they count on one another to share the booty they bring back from foraging expeditions with those that have been unsuccessful (DeNault and MacFarlane 1994; Wilkinson 1984). These animals practice "reciprocal altruism," which presumably requires them to keep a record of one another's behavior in order to shut out free riders.

In primates, a considerable part of the enlarged brain appears to be devoted to the social accounting and recognition tasks involved in life as a genuine social group (Dunbar 2003). Such a capacity is obviously an important stage in the path to full intentionality. For what counts for social life is not merely the specific kind of individual one interacts with but the concrete history of each particular animal—and what he or she did or didn't share in the past. It would be counterproductive to punish an individual who did not avoid sharing. Despite all that, an animal—bat or primate—could be fooled by an individual that manages to make itself indistinguishable from some other trustworthy conspecific. In the effort to distinguish cases of authentic reidentification from cases of imposture, only an act of "metacognition"—a second order judgment subjecting the claim of identity itself to further scrutiny—could make the relevant distinction explicit. It is not enough that the capacity for making such judgments should have been finely tuned by natural selection. Even if there is some sort of thought involved in the choice of whom to feed, that does not yet amount to the sort of reflective thought that is required for the animal to manifest rationality or irrationality.

In the grand sweep of evolution—in the long run of the species—it makes no difference whether sorting the cooperators from the free-riders is effected by thought or not. But for each individual, it is literally a matter of life and death. What matters to me, as an individual, is not what might be achieved or experienced by someone of my type, in the long run, as probabilities play themselves out in the fullness of time. What each individual holds to be of value, what each attempts to achieve, varies indefinitely from one person to another. That makes human values profoundly unlike the "value" served by an organ or an organism in virtue of its biological function. For the latter, from the point of view of evolutionary biology, is always the same.

This is not to deny the possible significance of the fact that certain animals in captivity, notably orcas and dolphins, have been reported to commit suicide by crashing into the walls of their enclosures. If true, these reports may well signal that these animals are driven to depression and self-destructive rage by the conditions of their captivity. Perhaps, indeed, the dolphins are able to weigh and discuss the merits of different courses of action. We don't know enough about dolphin "language" to say whether this is possible; but if it were, this would be an exception that proves the rule. If not, the case remains very different from the explicit adoption of a preference ranking for certain values above others.

Similarly, individuals or species can conflict or compete as rivals. But the logic of those conflicts dictates that the "value" served by each side is essentially the same in all cases: still and always, it is nothing but the vestigial value of replication.

Our own passions and our attitudes to our passions provide a clear illustration of the point I am belaboring here. Think of love and jealousy, for example. If the teleology served by my passions is not ultimately related to aims of my own, but determined rather, as I have been suggesting (and will shortly explain further), by the vestigial "goals" of gene replication, it is not surprising that my impulses should seem so indifferent to my own true interests and that the acts they motivate should so often prove counterproductive. For my passions may tend mostly to stop others' genes from gaining a reproductive advantage over my own, and that may turn out to be just as harmful to my own interests as to those of others.

It is often said that our deepest impulses elude rationality altogether. Insofar as that is true, it is mostly due to the fact that these allegedly deep motivations are not in some sense *my* motivations at all, but those of my genes, with which I may fail to share those specific interests. What, in the end, have my genes got to do with me? Why should I allow them to manipulate me so? Reflective thought is the only thing that makes it possible to raise, if not to answer, this metacognitive question. Without the power of thought, whatever I do and whatever happens to me, the ultimate *why* of my behavior can only reside *outside* me, in the vestigial teleology of my genes. The radically new domain that is opened up by the power of articulate thinking is the domain of multifarious individual values and the inner conflicts to which these now give rise: the newfound freedom, in other words, to be irrational.

This, perhaps, is the deep meaning of the biblical story about the forbidden fruit. The fruit concerned wasn't knowledge in general, but knowledge of good and evil; more precisely, the fruit of metacognition afforded by language. By means of the explicit resources of language, humans became conscious of themselves as individuals and empowered to invent new values transcending the vestigial values embedded in our biology.[17] In the rest of this book, I shall further explore the implications of the notion of an individual and its contrasts. I will begin, in the next chapter, by asking to what extent one can make sense of ascriptions of rationality to groups as well as to individuals.

Chapter 4 :: Rationality, Individual and Collective

*The gift of language was bestowed on man
that he might better conceal his thoughts.*

—Talleyrand

As we look at what people do and think, how do we assess their rationality? Our idea of what is rational and what is irrational can change radically, depending on how much context we take into account and whose point of view is in question. Three sorts of considerations can influence or even reverse our verdicts about what is rational: (a) frame and focus: how narrowly or broadly we identify the range of reasons taken into account; (b) point of view: whether we look at costs and benefits to the individual or to the community; and (c) temporal perspective: whether we take a long view or a short one.

The relevance of (a), framing, has already been discussed in chapter 2. The present chapter is devoted to (b), the analogies and connections between individual and collective rationality. I'll come to (c) in chapter 5, with a look at the difficulties of attempting to define standards of rationality through time.

::
4.1 The Individual and the Community

Recall the etiological account of objective teleology proposed in chapter 2: to ascribe a function to some process or organ in a system S is to claim that the process or organ in question exists in S now because of the contribution of just such processes or organs to the maintenance of a lineage of systems such as S. Thus the circulation of the blood is the function of the heart in virtue of the fact that the circulation of the blood has conferred fitness on countless generations of organisms and thereby resulted in their successful reproduction in the course of evolution. So the fulfillment of this function in the past has contributed to the existence of those hearts that are in vertebrates currently alive today. Such an account presupposes that it is possible to specify *for whom or for what* the exercise of the function is beneficial. The identity of the beneficiary, however, is not as obvious as might first appear.

You and I, as metazoan organisms, are typical individuals. My organs, the cells out of which I am made, the genes, the organelles that inhabit those cells—all of these seem to be working *for my benefit*. And yet, some mental mechanisms in my makeup seem to fly in the face of my best interests. My passions are notoriously liable, on occasion, to induce counterproductive behavior. Why should this be? Of course, they could just be malfunctioning. But a hypothesis worth considering is that such apparently self-defeating propensities are actually "designed" to serve not myself but other entities of which I am a part, or which are a part of me. It could be that those other interests are served by even my most irrational actions.

In the old Roman fable of the stomach and the limbs, by which Menenius convinced the people to end the first general strike on record, the stomach admonishes the other organs. It points out that that although the legs, teeth, and arms work for the stomach, they cannot live without it. Every organism is a community of interdependent parts: organs, assemblies, systems, circuits, cells. The great physiologist Claude Bernard regarded cells as the privileged unit. The body, he suggested, is "an entanglement of organs and systems, which exist not for the whole, and not even for themselves, but *for the sake of the cells*... thus the part depends on a whole which itself came together for the maintenance of the part."[1] But nearly every one of my cells, in turn, contains a quota of genes. And we shall shortly see why it is likely to be my genes, rather than my cells, whose interests are served by my irrational behavior. In the environment of evolutionary adaptation (EEA) of our ancestors, our propensity for behavior driven by passion tended, *on average*, to promote the propagation of the genes that have been handed down to us. If that is the case, my behavior will be likely to benefit those genes, even if it is definitely counterproductive from my own point of view.

Just as I can be regarded as a community of cells, I in turn belong to larger communities: a population, a family, the human species. Behavior that is irrational with respect to my own specific ends can benefit one of these larger entities. When it does, we call it "altruism," which misanthropists are as loath to ascribe to humans as animal lovers are keen to find in other animals. Witness the mythical lemmings who sacrifice their own lives for the good of their overpopulous species. In truth, the lemming story is only a fiction promoted by a Walt Disney movie. But something like it has been seriously claimed: the biologist Vero Wynne-Edwards (1962), for example, advanced the hypotheses that certain animals regulate their population size by imposing on themselves, as a group, a sort of spontaneous birth control. This could be interpreted as individual sacrifice for the sake of the group or species. Most biologists have been skeptical, but the notion of "group selection" has since regained some credence, as we shall shortly see.

Most premodern societies regard the tribe as more important than its individual members. The chief or monarch is the proverbial exception that proves the rule, for he is himself seen as the personification of the tribe. Indeed, the idea of the primacy of society has played a major role in certain modern communitarian ideologies.

From that point of view, sacrificial suicide can seem perfectly rational. It is "heroism." (At the opposite end of the scale, suicide could also seem rational from the point of view of a gene, if it is likely to result in the survival of more copies of that same gene.)

In sum, then, rationality hangs on point of view. For a single type of individual behavior, the ledger of profits and losses will depend on where you look: the society, the species, the organism, the cell, or the gene.

What Are Individuals?

We feel we can usually tell what might be in or against the interests of a person, or even of an animal. But our intuitions are less keen when we consider "individuals" of very different types. We need to look more closely at what it means to be an individual.

According to S. J. Gould, the vernacular notion of an individual boils down to "adequate stability between birth and death" (Gould 2002, 602). But he adds that if we want this notion to be useful in biology, we must specify certain supplementary criteria that explain the role that individuals play in the logic of natural selection. An individual worthy of the name must be a part of a *variable* population, be capable of *reproduction* of a sort that passes on *heritable* characters, and engage in *interactions* of a kind that affect its fitness. On the basis of these criteria, Gould sees biological individuals of different sorts at half a dozen levels of analysis. In addition to organisms, these include *cell lineages, demes* (relatively stable and isolated groups of members of a given species), and *clades* (both of the branches that are formed every time a lineage is divided into two distinct species) (Gould 2002, 612–13). Most notably, he claims that *species* can be treated as individuals extended in time and space.[2] In support of this view, Gould identifies several parallels between the properties of species and those of organisms. Species, like individuals, have their birth, consisting in the bifurcation of an existent lineage, their death, extinction, and their mode of asexual reproduction in the birth of a new lineage or "clade" (Gould 2002, 716–19). What particularly appeals to Gould is that this conception of species allows him to consider a kind of *group selection* acting directly on species, rather than on genes or individuals. Like other individuals at different levels, then, species would be *units of selection*.

Debates about units of selection have been strongly tainted with ideology. Group selection has sometimes been favored as a less reductionist alternative mechanism of evolution, in opposition to the *genic selectionism* championed by George C. Williams (1966) and Richard Dawkins (1976). These authors, in turn, rejected the classical Darwinian conception of selection acting on phenotypes, that is, organisms whose morphological or other characteristics make them more or less well adapted to their environment. A straightforward reason to reject this view is that there is, in fact, no *survival* of sexually reproducing organisms. Each organism breaks the mould: it is the product of a recombination of genes passed on from each parent to create an absolutely new individual, also doomed to die. Genes, by contrast, like single-celled organisms, are effectively immortal. Consequently, it is the gene, and not the ephemeral individual, that is the beneficiary of natural selection.

A gene or collection of genes has the "goal"—in the vestigial sense described above—of propagating itself. When we look at it this way, the commonsense view of the relationship between genes and individuals is inverted. It is naïve to think that my genes work for my benefit. On the contrary, I am the one who is an instrument of my genes.

In order to convince yourself of the truth of this paradoxical claim, recall the aetiological definition (P) of function given in chapter 2 above (p. 45):

(P) An existing element X has the direct proper function F if and only if:

1. X results from the reproduction of an antecedent element, *ancestral X*;
2. *Ancestral X* effected F in the past, in virtue of properties reproduced in X;
3. X exists now because *ancestral X* effected F.

To apply this formula to the relation between genes and organisms, we must make the following substitutions in the definition: "genes G" for X, and "made an essential contribution to the development of the human body" for "effected F." Thus we get (P'):

(P') d(1') *Genes G* result from the reproduction of *ancestral genes G*;
(2') *Ancestral genes G* have *made an essential contribution to the development of the human body* in the past in virtue of properties reproduced in *genes G*;
(3') *Genes G* exist because *ancestral genes G made an essential contribution to the development of the human body*.

We could conclude from formula (P') that contributing to the development of the human body is the *function* of our genes. Our genes would then be at our service, as we like to believe. But this raises a troubling question: *Who does this "we" refer to? To which human bodies does it pertain?* No two are the same. The development of a phenotype like mine cannot be the function of my genes, because the substitution of "this specific phenotype" in the formula does not yield a true statement. It must be false to say that ancestral genes contributed, in the past, to the development of organisms of the same type as I, because *my phenotype has never existed before me.*

That is why it is problematic to treat genes as having a function relative to an individual organism. However, if we invert the terms substituted for X, Y, and F in formula (P) to make (P″), all three of its propositions are true:

(P″) (1″) *Currently existing human bodies* are the product of the reproduction of *ancestral bodies;*
(2″) *ancestral human bodies* have carried out *the transmission of genes G* in the past in virtues of properties perpetuated in *genes G*;
(3″) *Currently existing human bodies* exist because *ancestral human bodies* have carried out *the transmission of genes G*.

According to (P″), bodies perform a function for genes, not the other way around. The reason for this is that genes, unlike phenotypes, preserve their identity as beneficiaries across vast stretches of time. No doubt their identity across time is not strictly maintained, since the very possibility of evolution rests on copying errors; but neither could evolution have taken place unless these errors were extremely rare. What is more, individual DNA molecules pass away like all material things. What subsists—as in the promise of Shakespeare's Sonnet LXV, *that in black ink my love may still shine bright*—is the *information* that we find in the lineage of sexual cells from generation to generation: "The gene is not the DNA molecule; it is the transmissible information coded in the DNA" (Williams 1992, 11). This is what justifies the claim that bodies fulfill their proper function in fostering the survival of genes. To put it more provocatively, *organisms are organs of their genes*. The converse does not hold, because individuals belonging to sexually reproducing species are never reproduced. Admittedly, the process we call "reproduction" suffices to maintain certain characteristics without which organisms would lack the fitness that enables them to transmit their genes. But organisms like ourselves are not capable of the sort of faithful reproduction that allows their copies to be reidentifiable across the ages.

A qualification is needed. We now know that, in addition to DNA, certain other structures get transmitted from one generation to another. These comprise several mechanisms that contribute to cell reproduction, and maybe even—and apparently without direct dependence on DNA—the entire "developmental system" that produces the organism.[3] While these discoveries are of great scientific interest, however, they don't affect the philosophical point I'm stressing here, which is that the goals of the individual are distinct from the teleology of the entities or structures, whatever they may be, that are the direct beneficiaries of natural selection. These last are *replicators*, which, like the "selfish genes" described in Dawkins's model, embody only what I've been calling a vestigial teleology, radically separate from the preferences and choices endorsed by individuals as such.

A startling illustration of this separation is the way that a simple gene, in the literal sense of a sequence of DNA, can exert its influence to the detriment of the individual in which it lives, even inside a gametic cell, thereby distorting the process of meiotic fission. This is known as *meiotic distortion*, and it results in a tendency for one allele to be favored over another at the time of meiosis. (Recall that alleles are the alternative forms of a gene at a certain locus in a chromosome; if there are two in a pair of chromosomes, their chances of making their way into the gamete that contributes to the zygote—the single cell that will give rise to an individual of the next generation—would normally be equal.) These distortions can persist, despite having a deleterious effect on the phenotypes that result from their presence in the genome (Dawkins 1982, 133).

An example of meiotic distortion is found in the house mouse, *mus musculus*. A haplotype (a complex of genes working together) of type *t* has the ability to fertilize an ovum more often than its rival of type +. But its victory is short lived because male homozygotes for this allele are sterile (Olds-Clarke 1997). So this results in a strange sort of conflict of interest between entities—the *t* allele and the organism of which it is a part—that are in any case bound to share the same fate.

The "genocentric" point of view endorsed here has been the target of much criticism. One particularly vigorous assault has come from Stephen J. Gould, who charges that the genocentric perspective involves a fallacy, which consists in confusing "bookkeeping" with causality. Gould writes:

> We may indeed...decide to keep track of an organism's success in selection by counting the relative representation of its genes in future generations. (In large part, we count at the gene level...because sexual organisms do not replicate faithfully and therefore cannot be traced as discrete entities across generations.) But this practical decision for counting does not deprive the organism of status as a causal agent.... The listing of accounts is bookkeeping—a vitally important subject in evolutionary biology, but not a form of causality. (Gould 2002, 633–34)

But what is meant exactly by "an organism's success in selection"? Why should I judge that the presence of my genes in future generations counts as a measure of *my* success? How could the success of my genes matter to me after I am dead? In truth, it is Gould and not Dawkins who is guilty of a logical error. The "units of selection" in which I'm interested are not claimed to be the sole *causal agents* of selection, but its primary *beneficiaries*. To identify those, it is precisely bookkeeping and not causality that counts, because identifying beneficiaries presupposes that we can identify entities with interests (albeit metaphorically). To be sure, it seems natural to say: *I am a beneficiary of the existence of my descendents because I have an interest in their existence*. But look again, and you'll see that the claim is circular. We don't notice this at first, because it's taken for granted that my own interests are bounded up with those of my progeny. Indeed, I *feel* concerned. But what does that show? Only that my emotions, too, are manipulated by my genes. Many people feel concern for the fate of "their" football club. Such feelings aren't self-justifying.

The problem disappears if we follow Dawkins in treating as potential beneficiaries of natural selection only those entities that subsist literally across generations, that is, genes as well as other "replicators" if they exist. The objections adduced by Gould and others against this point of view, which they deride as "reductionism," appear to be motivated by a curious fear of seeing human dignity "reduced" to nothing. To be nothing but the organs of our genes: does this not grossly demean us, who for so long have considered ourselves the instrument of an Almighty who made us in his image?

This fear is silly. Rather than seeing ourselves enslaved by the ends of the vestigial teleology of nature, we would do better to rejoice at not having to suffer the murderous terrors that have led men to kill one another, over many centuries, in obeisance to the obscure will of a tyrannical divinity. To be thus liberated doesn't mean I can't identify with the interests of my progeny. Neither does it degrade any other value that may be dear to me as an individual. On the contrary, it marks a clear distinction between the vestigial teleology of biology and the values that we ourselves assume. We need no biological justification to take interest in a stranger, a work of art, a moral ideal, or, for that matter, a snuff bottle collection.

⁂

4.2 The Design Illusion and the Emergence of Organization

When we contemplate natural phenomena involving populations, what we see often looks strikingly like the result of *collective design*. Yet the appearance of an encompassing plan or design is often nothing but an illusion, resulting from a large number of purely local interactions. The rhythmic waves of applause that we hear in a sports stadium and the structure of the Internet are just two of many examples of this phenomenon (Watts 2003). Let's look in a little more detail at two very different examples. The first, the "invisible hand" beloved of advocates of market-driven economics, actually illustrates certain problems inherent in group rationality. The second, "cellular automata," carries a number of broader lessons about the power of self-organization in collective phenomena. In the light of these two models, I'll then ask to what extent natural selection might also be regarded as producing an emergent hierarchical organization on the basis of local individual causes.

The Invisible Hand and the Tragedy of the Commons

Adam Smith uses the metaphor of the invisible hand in *The Wealth of Nations* to describe the manner in which a happy overall outcome can emerge through the combination of individual choices motivated purely by self-interest.[4] This invisible hand is often referred to as if it were a magical solution whose effects are always and only beneficial. And indeed, where it is allowed to determine the price of goods, the "invisible hand" of the free market accomplishes this with remarkable efficiency. But other collective outcomes of the general pursuit of self-interest are sometimes less benign. In a vast range of decision problems involving multiple rational agents pursuing their own interests, the collective as a whole will reach a stable equilibrium at a point where the value for all is far below what a benevolent dictator might have devised. The classic example is Garret Hardin's "tragedy of the commons" (Hardin 1968).

Hardin invites us to imagine a pasture available for use by every shepherd in the village. Being rational, each shepherd asks himself,

"What is the expected utility *for me* of adding an extra animal to my flock?" There would, of course, be some cost, consisting in increased deterioration of the common. But that disadvantage would be distributed among all who share it. The benefit to oneself, however, would far outweigh the cost, since it would be his alone. Each shepherd concludes that it is rational for him to add as many animals to his flock as possible. Because all pursue the same line of reasoning, the common is quickly degraded to the detriment of all. "The freedom of the common," Hardin concludes, "brings misery to all."

A similar road to ruin can be traced in a wide range of situations in which individuals exploit a limited common good. In a poor country, the decision to have an additional child may rest on the same logic. Each reasons that, whatever the others do, it would be better to have as many children as possible, all the while knowing full well that an expanded population is against everyone's interests. Here, too, we see a clear road to ruin. This problem is of the same game theoretical form as the *prisoner's dilemma* (PD), which I discussed in chapter 3. What the undesirable outcomes of these sorts of situations clearly show, then, is that the invisible hand is neither magical nor invariably beneficent. On the contrary, it seems to erect a formidable obstacle to the very notion of objective rationality as soon as it involves interactions between several agents whose interests are only partly consonant.

Cellular Automata

My second example of an apparently organized system that arises from exclusively local operations does not have the drawbacks entailed by any disreputable association with the prisoner's dilemma. To imagine a typical *cellular automaton*, picture a grid of squares or "cells" in a two-dimensional space. Each cell has two possible states—*on* or *off*. (There could be more than two states, provided that the total number of states is finite. That is, there must be a fixed number of states, without transitional states between them.) First of all, time is digital for a cellular automaton, meaning that successive moments unfold in discrete units, rather than in a continuous flow. Second, all change occurs exclusively in accordance with deterministic laws, uniformly applicable to all cells. These laws make reference only to the state of the cell itself and to that of its immediate neighbors. (By convention, we could regard each cell as having either four or eight neighbors.)

The state of each cell is updated at each discrete moment as defined by the universal clock. To give a concrete example, I paraphrase here the rules for the version John Conway dubbed "the Game of Life":

> *At each turn, check the state of each of the eight cells that encircle a given cell, C: How many are on? The number, N, of neighbors in the "on" condition determines the state of the cell in the next round, according to the following rules:*
> *(1) If C is off and N = 3, C turns on ("Birth" of C).*
> *(2) If (a) N = 2, or (b) N = 3 and C is on, C does not change ("Survival" of C).*
> *(3) In all other cases, C turns off ("Death" of C).* (Paraphrased from Gardner 1970).

The repeated application of these rules to certain initial configurations gives rise to an abundance of patterns, some of which exhibit a remarkable stability and coherence, as if they were conceived on purpose according to some global plan (see figure 4.1).

After experimenting with these a while, one cannot help being struck by the power of these few extremely simple rules. While some produce surprisingly orderly patterns, other rules bring forth *chaotic* complex structures that render the prediction of future states

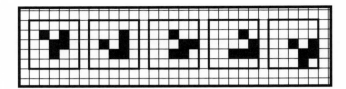

Figure 4.1. The Glider's Progress: In accordance with the rules in the text, the configuration cycles through four states and is reconstituted one square to the right and one square down. When programmed to loop rapidly through the cycle, the changing shape crosses the screen, looking like a coherent gliding form.

practically impossible.[5] In his enthusiasm for models of this kind, Steven Wolfram goes so far as to speculate that we may more successfully explain the complex phenomena of nature if we abandon the traditional approach of finding the differential equations that describe continuous processes. Instead, we should look to cellular automata in our search for understanding of the way the physical world behaves. Among the configurations of discrete elements whose evolution in time is governed by simple rules of purely local application, we might find the models to illuminate physics and biology (Wolfram 2003).

While such extreme claims may be overblown, we can draw three lessons from this brief characterization of cellular automata. The first is purely abstract, but it is of great importance in regard to certain critiques of Darwinism generated by the partisans of so-called Intelligent Design theory. The master argument of this line of criticism rests on the claim that the level of complexity manifest in the living world could not possibly result from selection working on random modifications.[6] That claim looks feeble in the light of the fact that we can obtain chaotic patterns of arbitrary complexity from a small number of childishly simple rules. It's difficult to assess the full force of this counterclaim made on the basis of cellular automata, however, because it must be conceded that those models are very abstract and lie at some distance from concrete biological realities.

The second moral to be drawn from these toys is that we should not accept too readily the conjecture that our universe is structured according to continuous mathematical quantities. It might well be, in the end, that the ancient Greek atomists were correct in positing that the universe's fundamental structure is granular. The discontinuous universes of cellular automata give rise to structures of such richness that it is not implausible to surmise that even a fundamentally deterministic universe of this sort could account for the apparently infinite abundance of observable phenomena, together with the practical impossibility of predicting the future with certainty.

Third, we can see clearly that these structures lend themselves directly to the construction of digital representations, which in turn facilitate a broad range of high-fidelity reproductions. I have already emphasized the importance of an extremely high, but not quite perfect, reproductive fidelity in the process of evolution. Cellular automata lend additional plausibility, then, to the idea that natural selection might fashion the appearance of smooth change out of ultimately discontinuous factors.

Natural Selection

Like the two models just described—the invisible hand and cellular automata—natural selection illustrates the effect that local causation can have on global structures. Each individual event contributes to an effect that seems to manifest a global tendency or direction. But for reasons I have tried to explain, we cannot clearly assess the positive or negative effects of the invisible hand—the favorable or unfavorable character of equilibria attained—unless we specify the individuals whose interests are at stake. In certain cases, such as meiotic distortion, natural selection can produce results that are far from optimal, no less for the entities that are the direct cause of the effect (such as the gene that results in meiotic distortion) than for the organisms of which the gene is a member—whether we have in mind the organism, the lineage, or the species.

These cases present a curious analogy with the *minimal rationality* that subsists even in the most egregious examples of irrationality. Recall Andrea Yates, whose case was described in chapter 2. Despite the gross irrationality of her project of killing her children, she went about the task in thoroughly methodical fashion. In much the same way, the immediate effect of allele t is to secure for itself an advantage in a minimal setting by enhancing its chances of out-competing its allele. However, this effect sometimes dooms the organism, and hence the gene itself. Indeed, this "sometimes" can be quantified, insofar as it depends on the frequency of the gene in the population. (The notion of "frequency-dependent fitness" will be explained in section 4.4 below, in connection with the question of how prudent it is to lie.) The positive effect of meiotic distortion, in virtue of which it supports itself, is thus thwarted by selection at other levels. This explains why the effect remains weak.

4.3 Does It Make Sense to Speak of Group Rationality?

The philosopher Georges Canguilhem once wrote that social organization "is the solution to the problem of converting competition into compatibility" (Canguilhem 1994, 330). In fact, that conversion is often impossible. The simplest illustration of this impossibility, known as the "voter paradox," was discovered in

1785 by Marie-Jean-Antoine Nicolas Caritat, Marquis de Condorcet, who showed that we cannot even make a coherent generalization about the transition from individual preferences to a collective preference (Condorcet [1785] 1986). Imagine three electors who need to choose among three candidates, A, B, and C. It seems reasonable to expect that we could derive the group's preference from the individual preferences of its members, on the basis of a majority vote. Two out of three votes would win out. But suppose that the three electors order their preferences as follows:

1. $A > B > C$.
2. $B > C > A$.
3. $C > A > B$.

It is easy to see that the resulting majority order of preference is circular. The votes are two to one in favor of A over B, two to one for B over C, and two to one for C over A. The group preference, then, seems to be $A > B > C > A$. But in assessing individual decisions on the basis of preference, it seems indispensable to abide by a principle of *transitivity*, requiring that, in preference orderings as in the natural numbers, $[A > B \text{ and } B > C]$ implies $[A > C]$. On the basis of that principle, the collective order of preferences arrived at in the voting paradox is incoherent.

Nobel Prize–winning economist Kenneth Arrow generalized this result by showing that if we satisfy the condition of *independence of irrelevant alternatives*, no function will lead from a collection of individual preferences to a coherent collective preference (Arrow 1964). The independence of irrelevant alternatives is best illustrated by a comical anecdote in which it is violated. The philosopher Sidney Morgenbesser was offered a choice of apple pie or blueberry pie for dessert. He chose the apple pie. But when the waitress returned a moment later to add that strawberry pie was also available, Morgenbesser quipped: "In that case I'll take blueberry." The comic effect of ignoring the rule underlines its plausibility. We will see in chapter 5 that it's actually quite hard to say just what counts as "irrelevance"; for that reason, the norm of transitivity is not easy to apply. Still, we infringe this norm at our peril, since an infraction could be exploited as a perpetual "money pump." (*Since you prefer B to C, you'll pay a little more for B. And since A is better than B, you'll pay a little more for A. But you also think C better than A, so that should be worth still more....*) Provided an agent is willing to pay to have what she prefers, we could bleed payments from her indefinitely. This would be inconvenient, to say the least, for a society whose collective order of preferences fell into such a circular pattern.

By and large, competition and cooperation go together, like an old couple whose long relationship is fraught but steady. The advent of multicellular organisms and that of sexual reproduction are just two of the "major transitions of evolution" singled out by Maynard Smith and Szathmáry (1999). Both bear witness to the fact that we encounter, throughout the living world, not only undiluted competition, but also cooperation as a strategy of competition. One important aspect of this dialectic plays out between egoism and altruism: I will come to this presently. But first, let us look at another prominent example of the ambivalent relationship between the individual and the group. Once again, this one illustrates the parallelism between an intentional phenomenon and a purely biological one in which intentionality plays no role. I refer to the question of the rationality of lying and truth telling.

4.4 The Rationality of Lying

Immanuel Kant notoriously challenged common sense by declaring an absolute prohibition against lying. Lying violates the categorical imperative, which demands that I be able consistently to will that everyone always act as I do. If I lie, according to Kant, "from the moment that it is established as a universal law, my maxim will necessarily destroy itself." It would destroy itself because lying is possible only if we can expect to be believed. Without a norm of truthfulness, people would have no reason to believe one another, and lying would be pointless (Kant 1998).

Kant's claim has the lofty irrelevance of many an *a priori* pronouncement. As one critic nicely put it, "None of us is in the least discouraged from applying a general rule by the thought that there are exceptions" (Livet 2002, 216). The moral issue does not concern me here, so I won't enter into the debate about whether Kant was right to ignore that obvious sociological fact. But I do wish to argue that one is wrong to ignore it from the point of view of rationality. To see why, let's first compare lying with its purely biological counterpart, which is *animal mimicry*.

To see the significance of mimicry, we must grasp the concept of an *evolutionarily stable strategy* (ESS). This is a phenomenon of prime importance for the relationship between individual strategy and collective solutions to problems of cooperation and competition. The key fact here is that the fitness of a given trait sometimes *depends on the frequency of this trait itself* (Maynard Smith 1984) (Dawkins 1976). Once again, this is easy to understand in economic terms: if no one else is making shoes, setting up a shoe factory is likely to be profitable. If there is a glut, you had better find another line of work.

The essential feature of an ESS is that it is *stable*: there is no reason to expect that the precise point of equilibrium will be especially advantageous either for the individual or for the group of which it is a part. The "sex ratio" in mammals (the proportion of males to females) provides a handy illustration of the mechanism that sets an equilibrium of this sort. Sexual reproduction in itself does not demand a sex ratio of one to one. Indeed, if nature were the benevolent mother she is so curiously often supposed to be, she should surely be more sparing in her allotment of males. Their number—not to mention their very existence, when we recall that some "parthenogenetic" species do without males altogether—is not necessarily optimal for the species. A majority of females would more economically allow for the renewal of the population, while also providing for gene repair and genetic diversity—two functions widely credited with compensating for the disadvantages of sex.[7] Nevertheless, the proportion of the sexes at the moment of conception remains obstinately around 51 percent males to 49 percent females. That ratio—given the higher mortality rate among male fetuses and infants—secures an equal ratio at the age of maximum probability of reproduction. The sex ratio therefore illustrates the purely mechanical logic of ESS: as soon as one sex acquires a numerical advantage, the other's chances of spreading its genes (including any propensity to produce children of the minority sex) increase in the next generation. The ESS is thus quickly reestablished.[8]

Let us return now to truth and lying. If we think of them as alternative strategies, we can expect truth-telling and lying to settle into an ESS. Elliott Sober has shown how this can be seen in the pure, intention-free domain of animal mimicry (Sober 1994). Here's a well-known example: the Viceroy butterfly is nonpoisonous, but it looks like the highly toxic Monarch. In this way, it wards off potential predators that have been selected for their disinclination to consume Monarchs (or perhaps—it makes no difference to the logic of the case—learned to avoid them after a toxic nibble). As Sober points out, the proportion of Monarchs to Viceroys will settle at a distance from either of the two symmetrical extreme cases. If no one imitates the Monarch's coloration—if everyone speaks the truth—any eventual mutant whose coloration approaches that of the poisonous Monarch will prosper in its "deception." At the opposite extreme, if we imagine an ecological disaster in which no Monarchs but only Viceroys survive, the latter's disguise will quickly lose its protective power.

Notice that the same goes for the butterflies' predators. The few blue jays that would earlier have put themselves at risk by eating either Monarchs or Viceroys are analogous to the skeptical individuals whom liars find it impossible to exploit. Their skepticism will be vindicated, and they will now take advantage of a food source neglected by their more prudent conspecifics. In this new situation, the alimentarily adventurous blue jays would be particularly likely to proliferate, but only while they are few. When their numbers reach the threshold at which most blue jays are equally well informed, the competition levels the field once again. The advantage that they enjoy, then, like that of the Viceroy butterflies, depends on their proportion in a population. Both the success of the deceiver, and that of the skeptic who guards himself against being taken in, are frequency dependent. The point of equilibrium, the ESS, depends on several parameters, including, of course, the number of predators and the risk that they run; but most notably, for any set of values of those parameters, it depends on the proportion or frequency of Viceroys and Monarchs.

The pattern is quite similar in the case of intentional lying and truth-telling, except for one important difference. In an intentional act, we can usually distinguish the components of belief and desire that have influenced a given decision. The rationality of a belief and that of a desire are not necessarily the same as the rationality of the behavior that follows from them. In the case of animal mimicry, by contrast, the parameters are impossible to untangle. We can observe that the blue jay avoids eating the Viceroy butterfly, but this behavior is compatible with a whole continuum of possible explanations. For example, we could attribute it to a belief that there is an 80 percent chance of the butterfly's being poisonous, coupled with a modest desire to avoid a mildly disagreeable experience. Or we could suppose that the jay only believes there is a 5 percent chance of danger but that it is convinced that eating a poisonous butterfly will cost it its life. Which is the correct explanation? Despite the notorious ingenuity of researchers in animal cognition, we are unlikely to discover the answer for the simple reason that we can't get the blue jays to tell us. Here again we encounter the phenomenon remarked upon in chapter 3: only explicit language can set fixed values to the parameters of desire and belief, in such a way as to justify a verdict of "rational" or "irrational."

4.5 The Solution to the Problem of Altruism

As a theoretical question, altruism is to biology what evil is to theology: an in-house problem designed to keep the firm busy. Without the hypothesis of an all-powerful and benevolent deity, there is no "problem of evil." In the same way, without the hypothesis of natural selection there is no "problem of altruism." In each case, the "problem" arises from the fact that the theory has apparently defined a manifestly real phenomenon (evil, or altruism) out of existence. The theory then struggles either to resolve or to bypass the problem it has created.

For common sense, an altruistic act is one *motivated* by concern for the other rather than by the agent's own interests. But in biology, the notion of altruism bears only indirectly on motivation. An act that is altruistic in the biological sense is defined as one that diminishes the fitness of an agent while benefiting one or more other individuals. If there is a gene favoring such behavior, it seems obvious at first sight that it must leave fewer copies of itself than an allele that favors objectively selfish behavior. Sooner or later, the genes that favor selfishness will have eliminated those that favor altruism. And yet, altruistic behavior exists. Therein lies the problem.

Sociobiologists have offered three sorts of explanations: *kin selection*, *reciprocal altruism*, and, most recently, what I will call *Simpson selection*.

Kin selection builds on the observation that I share a certain proportion of my genes with my close relatives. This fact could explain why I am disposed to sacrifice myself for them to an extent proportional to our degree of kinship. This hypothesis sharply brings out the need, discussed above, to identify the *beneficiary* of selection. The majority of parents take it as obvious that the interests of their children can be conflated with their own. For whom do we work so hard, after all, if not for our children? Self-sacrifice in favor of one's offspring seems to require no further explanation. But philosophy is nothing if not the bizarre compulsion to chase explanations where none seem required. And the fact is that if we persist in asking why it seems so obvious that "my children are a part of me," we hit a blank wall. The real explanation for the interest I have in my children must appeal to a teleology *in which neither my children nor I play an essential role*. From my point of view as an individual, my attitude is nothing but brute fact: I am just "wired that way." And the genes responsible for this wiring, which makes this attitude "obvious," owe their existence in my body in part to the fact that this same attitude conferred on my ancestors the benefit of their parents' protection. From the gene's eye view, we are, in Dawkins' striking image, nothing but *vehicles* charged with safeguarding and transmitting these genes. But there is a determinate probability that the genes I bear also reside in another body, according to the degree of our strictly biological kinship.[9] So it makes sense after all that I should be motivated—without, to belabor the point, necessarily understanding why—to "risk my life for two siblings, or eight cousins."[10]

Now if I want my behavior to conform precisely to this calculus, I must know the exact degree of my kinship with others. Since it is unlikely that my knowledge or reactions could be so minutely regulated, the strategy most likely to spread would be one that more vaguely favors the whole clan to the detriment of the individual. At least, this would be so providing two conditions are satisfied. The first is that the clan be sufficiently restricted in size, and its genome relatively isolated, so that the genes in any one individual should be widely distributed amongst the community; the second is that the group's cohesion should be sufficiently long lasting for the effect to be felt.

Although it seems unlikely that both these two conditions should be met very often, they may have arisen a sufficient number of times to promote the mechanism of kin selection, and thus favor objectively altruistic acts. The mechanism would be preserved because it benefits the genes, not the individual altruists. In this way, "selfish genes" can be reconciled with a certain sort of "group selection."

The second model proposed to explain the existence of altruism is *reciprocal altruism*. This presupposes a capacity to recognize individuals who have contributed to my well-being in the past. This mechanism doesn't require that each act of kindness be specifically rewarded, but it will be effective to the extent that most individuals are capable of telling freeloaders apart from cooperators. That will provide motivation for avoiding the consequences of a bad reputation. I have already had occasion to mention vampire bats, which may have the capacity to keep track of the past conduct of individual members of the colony (Wilkinson 1984). If so, the genes benefit from the process of natural selection that have resulted in that species' capacity to keep track of individuals. But in this case the benefit rests on the tracking of past benefits being sufficiently widespread in the colony as a whole to make food sharing worthwhile. And here, too, the mechanism will work effectively only in colonies small enough to allow mutual recognition among its individual members.

The third model offered to explain group selection of altruism owes its name to *Simpson's paradox*[11] which designates a range of situations in which the proportion of individuals that meet a certain condition in a large group is reversed in each of the smaller groups of which it is comprised. The following example provides an illustration.

> A university was accused of sexual discrimination, on the ground that the proportion of men applicants admitted was 14 percent while the proportion of women applicants admitted was only 12 percent. Upon investigation, it turned out that in each faculty (suppose there are only two: "Humanities" and "Electronics") a higher proportion of women than men were admitted. Overall, though, the proportion of males admitted remains higher.

Table 4.1 sets out one possible combination of statistics that exhibits these paradoxical features. Note that a disproportionate number of men apply to the faculty of Electronics, which has a much higher rate of acceptance. It is also the case that there are more women applicants overall. It is by virtue of these two conditions that there can be a lower proportion of women accepted overall, even though the opposite is true of each faculty considered on its own.

According to Sober and Wilson (1998), this mechanism could account for real cases of group selection, which would displace the gene from its monopoly on the title of beneficiary of selection. For that to work, we must assume a population divided into subgroups that remain relatively isolated from one another for a sufficient length of time. Some of these subgroups contain very few altruists,

:: Table 4.1. Simpson's Paradox

Faculty	Applied		Admitted				Majority
	Male	Female	Male	% Male	Female	% Female	
Humanities	10	100	1	10%	11	11%	F > M
Electronics	40	10	6	15%	2	20%	F > M
Total	50	110	7	14%	13	12%	M > F

The rightmost cells summarize the data on applications and admissions, showing that in each faculty, the proportion of female applicants admitted is higher than of males, despite the fact that, nevertheless, a greater proportion of male applicants are admitted overall.

while others have more. On average, the altruists' sacrifices have two results. First, altruistic individuals would, by definition, leave fewer descendants than the selfish individuals *in any given subgroup*. Second, all members of groups in which altruists chance to predominate, whether selfish or altruist, would enjoy a higher fitness in comparison with members of subgroups consisting of mostly selfish members. The first result means that an altruist subgroup is open to being "invaded" by selfish individuals, whether by mutation or because there are one or two such individuals in the group from the beginning. The proportion of altruists in each subgroup would go down with time, ultimately reaching the point at which they have disappeared entirely, and with them the benefits they conferred upon the group. But in virtue of the second result, as long as a subgroup consists mostly or exclusively of altruists, its members will enjoy the net benefits of their altruistic behavior and will become more numerous. The overall fitness of subgroups invaded by selfish individuals is lower than that of subgroups that remain dominated by altruists. Hence it is possible (depending on the values of the parameters in play) that altruists could come to be in the majority overall, as a result of the superior fitness of subgroups of altruists, despite the fact that altruists will get squeezed out of most subgroups taken in isolation. Such a mechanism would give rise to group selection, insofar as it would benefit from the greater fitness that individuals acquire by virtue of their belonging to certain groups.

For an illustration of this last model, we can look back at the fact, discussed in chapter 3, that some ants continue to forage at random even when all the others have put a winning strategy into play by following existing pheromone tracks. The behavior of the random foragers who forgo the trodden paths is objectively altruistic because they run a higher risk than the others of not getting fed. But their existence is indispensable to the survival of the colony when a food source dries up: only they can bail the others out by discovering new resources. In the light of the model of Simpson selection, we might surmise that colonies with no altruistic explorers would have died out, preserving in all viable colonies a minimal presence of altruistic foragers. In this case, however, kin selection is certain to be involved as well, since the ants in a given colony tend to be closely related.

The models of group selection I have described are plausible, at least in theory. But in expounding them I have glossed over an obscurity in the very idea of "group selection." Taken literally, this term raises two questions.

First, what exactly *is* a group? What are the *identity criteria* for a group? Surely a group can't simply be defined as the sum of the individuals that make it up, any more than an organism can be identified as a particular set of cells. A group might remain the same while its members are replaced, just as a person remains the same despite a regular turnover of individual cells in our bodies.

Nor can the group's identity depend on its geographical location or its ecological niche, for otherwise it would be impossible by definition for one group to replace another group at the same site. Conversely, we can imagine the *cultural* disappearance of a group that would leave intact the identity of its genome, as when a "conquering group forces the conquered group to adopt its norms and institutions" (Fehr and Fischbacher 2003, 790). In such a case, the *group* could disappear without its *members* being any the worse for it.

Second, how should we define the *progeny of a group*? Does the term apply to those individuals who are descended from members of the group, without further consideration of whether they still form a single cohesive group? Or can a group have as progeny only another group, on the model, perhaps, of a new insect colony established by the emigration of a queen?

According to the answers one favors to these two questions, the term *group selection* could have very different meanings, and there seems to be little rationale for preferring one answer over another, unless one appeals to precisely what group selection is supposed to supersede, which is the "bookkeeping" of the genocentric perspective.

::

4.6 The Role of Sanctions

I have so far neglected the part played by *culture* in the fitness of a group or of the individuals that make it up. The role of culture is of utmost importance, because the three mechanisms cited above—kin selection, reciprocity, and group selection—do not sufficiently explain the presence of altruistic behavior in *large* groups on the scale of human societies. Modern urban populations are quite simply too large for any of those mechanisms to get much of a grip. My degree of kinship with a random stranger is low; any benefit my genes might derive from my altruism, therefore, will be so diluted as to be negligible. Reciprocal altruism is effective only among individuals that can remember one another. As for the specific mechanism of group selection derived from the Simpson phenomenon, it involves the well-timed dance of dissolution and reconstitution of subgroups in a large but still cohesive population. It is not clear that the right conditions would be met often enough to have any significant influence on human evolution.

We could make do with the hypothesis that, when our altruistic actions benefit strangers, these strangers benefit from a sort of erroneous extension of our propensities from true kin to the few million inhabitants of our city that are not of our tribe. True altruists would then be individuals whose power of kin discrimination has been weakened or lost, either by mutation or owing to changed circumstances.

An intriguing illustration of how this sort of mechanism might work is provided by a study of Israeli kibbutzim by Edward Westermarck (1922), and more recently confirmed by a study of a Taiwanese culture in which adopted daughters were raised with the sons they were intended to marry. Westermarck noted that adults raised together from childhood in the same kibbutz seem to feel little sexual attraction for one another. In this case as in the case of altruism, a trait has been selected for in an environment of evolutionary adaptation (EEA). Avoidance of sex with those perceived as siblings served the function of avoiding the genetic consequences of incest. And in that environment, proximity was a good enough marker of kinship. When the situation differs from that original EEA, and the children raised together are not brothers and sisters, the same behavior will be triggered by old cues. But these same cues, based on close proximity, have now lost their connection to the debilitating consequences of incest. The behavior in question will therefore no longer have the same genetic consequences. It is now merely the consequence of an error, a malfunction of the kin identification function in the service of incest avoidance.

As a vestigial function, it will no longer be selected for. But a function that is not reinforced by selective pressure is in danger of gradual extinction. It would be worrying to think that human altruism ultimately rests on a sort of objective malfunction. Once more, it appears that it might be dangerous to place too much trust in the ability of our natural dispositions to keep us happy, healthy, and moral.

The hypothesis that, strictly speaking, altruism might be in part the result of a biological malfunction may seem implausible. But it should serve as one more reminder of the radical separation between biological "value" and genuine human value, on which I have been insisting throughout this book. Furthermore, the error hypothesis may seem less implausible when we recall that a trait can be perpetuated without either going to fixation or disappearing altogether. Genuine altruistic sacrifice, in people as in foraging ants, is actually rather rare, so there is perhaps less to explain than is often assumed. Furthermore, many who sacrifice themselves for a cause or for their country, like soldiers or suicide terrorists, are surrounded by a great number of militants who perhaps become a *de facto* emotional "family" (Pape 2003). This may be regarded as a psychological confirmation of the error hypothesis.

Still, it may be that some supplementary mechanism exists, which reinforces other-directed behavior in a society too populous to admit of a universal mutualism. A number of researchers have recently promoted a version of this hypothesis, supposing a conformist predisposition toward *imitation*, consolidated in its turn by a will not only to punish infractions but also to *punish those who neglect to punish the infractions of others*. Fehr and Fischbacher

(2003) have studied what would happen over a span of two thousand "generations" of the PD game, iterated with multiple participants in groups ranging from 2 to 512 members. If infractions are not punished, cooperation falls to zero by the sixteenth turn. By punishing violators, cooperation is maintained a bit longer, but it still disappears completely in groups of 128 members or more. But when those who do not punish the infractions of others are themselves punished, Fehr and Fischbacher found that a stable level of cooperation was maintained all the way up into the largest group tested in the experiment, with 512 members. We are still far from tens of millions here, but the simulation is highly suggestive. The authors conclude by formulating four essential conditions for the general dominance of societal norms over the immediate interests of individuals: (i) individuals must be able to earn a *reputation* that can either serve or harm them in the future; (ii) they must have a strong innate tendency to *imitate the majority*; (iii) infractions must be subject to *punishment*; and above all, (iv) *failure to punish infractions must be punished*. These four factors together are powerful enough to turn the expected utility of selfishness from positive to negative. That would suffice to eliminate genetic selection in favor of individual selfishness. The upshot is a cultural-genetic coevolution that reinforces the influence of the group on the individual.[12]

This influence does not exclusively favor the nicest forms of altruistic behavior, as Boyd and Richerson point out. On the contrary, it provides just as good an explanation for tyranny and fascism, which should, come to think of it, be regarded as just rather special forms of altruism. All the same, group influence provides the best solution to date to the problem of the *possibility* of altruism raised by the theory of natural selection.

Over the course of this and previous chapters, I have evoked several parallels between the results of individual rationality and those of natural selection acting on large numbers. We have seen that extending the domain of rationality to groups poses a number of problems. But the problem of extending rationality from individual to collective can return, boomerang-like, to plague the very notion of individual rationality itself. For individuals are societies of cells, harboring genes whose objective teleology sometimes pits one against the other, or against other stable entities—cells, organisms, species, social groups—that can be identified at other levels of organization. Not only social situations in the usual sense, but biological interactions of various sorts, can exhibit the structure of a prisoner's dilemma.

Once more, then, the rationality of a behavior depends on the entity, be it individual or collective, on which we focus our interest. We have seen how this works in a number of different cases. An additional sort of relativity arises in the case of ESS's, in which a differential frequency can be of systematic benefit to the trait that finds itself in the minority. We are mistaken, then, if we assume that a genetic trait of any sort must either present an advantage or prove harmful in a given natural environment—unless we remember that other members of a group or species are part of that environment. If traits were intrinsically beneficial or harmful in a given environment, we could expect each trait either to "go to fixation"—that is, to become universally established—or to be completely eliminated over the long run: individuals deprived of an advantage will disappear, while those who are blessed with it will flourish. But things turn out differently with traits whose advantage is frequency dependent. No allele will be definitively eliminated, and so genetic diversity will be favored. To the extent that an ESS favors the survival of the group in a given environment, barring excessively abrupt changes, it will have a net favorable effect on the population across generations. On the other hand, it will often force populations to remain confined in a situation that is well removed from what a rational agent might have chosen, similar to (or worse than) the second-best option equilibrium in the prisoner's dilemma situations in which both participants act in a "prudent" way.

But what, once again, makes a "rational" or an "irrational" agent? It turns out to be quite complicated to say particularly because, as I have already hinted, the very same behavior can count as both at once. To explore some of those complications will be my task in chapter 5.

Chapter 5 ⁝⁝ Irrationality

All great things come to us by way of madness, gift of the gods.

—Plato, *Phaedrus*

All success is partly due to luck—if only because it always depends on circumstances beyond the agent's control. Rationality increases my chances, but it provides no guarantee. In an ideal world, where any experiment could be repeated as often as needed, rationality would pay in the long run. In the world as it is, however, the actual rate of success doesn't always match the probabilities. We would judge it suspect if the same number came up on one hundred consecutive throws of a die. But in the absence of such an extreme anomaly, we wouldn't let the odds on the next throw be affected by previous results. We would trust the shape of the die, which promises odds of one in six.

Things are a bit different on the timescale of evolution. Natural selection has plenty of time to zero in on the actual ratios predicted by probabilities and degrees of fitness, and probability can be interpreted literally as referring to the proportion of different alleles expected in the members of future generations. The gains and losses at stake in the "choices" embodied in the current set of genes will be tallied up by the future distribution of those same genes. In the meantime, however, the rules of the game may well have changed, so that the race to achieve the expected ratios is like the story of the tortoise that Achilles will never catch, or of the Red Queen who runs only to remain in the same place.[1] In the end, then, biological fitness has no greater chance of attaining perfection than does rational behavior.

If rationality offers no guarantees, how much worse is it to be *irrational*?

5.1 Strategic Irrationality and Epistemic Irrationality

We act, and we think. Depending on which is in question, we can speak of two modes of rationality. The *strategic* mode applies to action and to means and ends desired; the *epistemic* mode applies to belief. A handy way to distinguish the two is by their *direction of fit*: with belief, success is a matter of fitting representation to the objective world; with action, it is the world that must be made to fit a goal.[2] I will explain presently why we should also consider a third mode, the *axiological* mode, which governs the rationality of emotions and is concerned with values. But let us keep to the first two modes for now. Strategic rationality is the object of study in decision and game theory. Epistemic rationality is the concern of logic and epistemology. Both aim to maximize the chances of success in their respective domains. For the first, what counts as success varies according to the act envisaged and is relative to the agent's choices and desires. In the second, there are two separate goals: to attain truth and to avoid falsehood.[3]

The notion of success is normative: in all instances of irrationality, some norm is violated, something is done or believed that in some sense ought not to be done or believed. But what sort of norms are these? A naturalistic perspective rules out the hypothesis that norms are handed down by a superhuman authority. But if human beings set them up, what makes them right? Must we fall back on the democratic criterion of majority rule? This idea is popular in some circles, which regard the so-called norms of rationality as nothing but an arbitrary power play. Once we reject the divine provenance of norms, they are bound to be contested.

I need to specify what I mean here by "norm." Ruwen Ogien and Christine Tappolet have elaborated a fine critique of the notion of a norm, which they urge us not to confuse with that of value (Ogien and Tappolet, forthcoming). A value might derive from subjective preferences or desires; it would then admit of degrees, reflecting the vicissitudes of preferences and the variable intensity of desires. Moreover, we can accord greater or lesser value to the existence of a norm: "Brush your teeth" seems to be a more important norm than "Don't kill people." Values, such as the importance to well-being of secure property, can be used to justify a norm, such as "Don't steal." In themselves, however, norms are not values and do not admit of degrees. It may seem that they do because we can measure the gravity of the sanctions warranted by a specific violation or the intensity of guilty feelings to which it gives rise. One can also rate the importance of different domains to which norms relate—norms of etiquette, for example, are generally rated less important than norms of morality. But the fact that sanctions, sentiments, and importance all admit of degrees does not imply that the same is true for norms themselves. A norm, in the strict sense, prescribes what is required or forbidden. It can be followed or infringed, and nothing else.

So when I speak of rationality as a norm, am I speaking of norms in the strict sense, or of values? I intend to sidestep that question by claiming that rationality is at the very least *normative*. The sense of that word is broader than the strict sense of *norm*. Normativity, I suggest, is involved as soon as it makes sense to complain about or *criticize* something done or believed. Ogien's distinction remains important because some reproaches are categorical, while others are relative or nuanced. I will speak then of *Ogien's criterion*: if we are dealing with a real norm, the obligation to conform does not admit of degrees. If rationality is not achieved until some value is maximized, then success in that area does not admit of degrees. Either the maximum is attained or it is not. In that case, rationality will be a matter of norms. By contrast, if we can be more or less rational, then rationality will not be considered a norm in the strict sense; but it will still be normative in the more general sense.

In the specific case of belief, it seems clear that truth is a full-fledged norm: either a belief is true or it is not. Doubtless some clever reader will object that this assertion is only *more or less true*. I brashly reject this counterexample by adding the simple qualification: *properly speaking*. *Truth proper* does not admit of degrees. Degrees do come into play, and we can justify using the expression "more or less true," to the extent that we are interested in other virtues a belief may possess: pertinence, sincerity, precision, universality, all of which do admit of degrees. (And more generally a goal can be approached *more or less closely* without being attained.) Furthermore, one can be more or less confident in one's belief. But none of this means that there can be degrees of truth.

Sometimes we *think something likely* and treat it as true for practical purposes. But sometimes we just believe it, without qualification. I will call the latter kind of belief, the sort that attaches to a proposition without qualification or quantification, "absolute belief."[4] Absolute belief is a *discrete* state, which can figure in a *digital* system of representation: either I believe that p, or I do not (and the same applies for the negation of p). There is no middle ground. An absolute belief's criterion of success is simply the truth of its object.

Armed with this terminology, I propose to stipulate two criteria to characterize the sort of normativity governing any categorically rational state or act R. First, in believing or doing R, the agent *assumes the risk of meeting with a specific sort of criticism*. Second, R's rationality is *analytically tied to the conditions of success for R*.

The norm of rationality for absolute belief, for example, is analytically tied to the criterion of success for belief, which is truth. These two criteria also define the stricter notion of a norm. Whatever the advantages of self-deception, a false belief is always a failure according to the intrinsic criterion of success for belief.

5.2 Strategic Belief and Pascal's Wager

There have been champions of an alternative point of view on the rationality of belief. A well-known example is Blaise Pascal's wager on faith. This wager rests on a calculation: even if the probability that God exists is infinitesimal, choosing faith is nevertheless always more rational than choosing atheism because it has a higher expected utility.

We can represent this calculation as an application of Bayesian decision theory. The principle behind this theory is simple and compelling, and it applies to all rational wagers. Recall the formula from chapter 3:

$$V = \sum_{i=1}^{n} (p_i \bullet v_i)$$

The "expected value," V, of an action is equal to the sum of the values of each possible consequence, v_i, weighted according to its probability, p_i. In a Bayesian context, probability is subjective. It represents a *degree of belief*, as opposed to absolute belief. If I place a bet on six in one throw of a six-sided die, for example, the rational degree of belief that I will win is 1/6. Assuming the bet is fair, then, I should get six times my outlay if the six comes up.[5]

So we can we can express Pascal's wager in Bayesian terms as follows.[6] The value of eternal bliss is infinite. Let us also assign a value of 1,000 units to a life of pleasure and a value of negative 1,000 to an ascetic life. (The actual values are arbitrary: all that matters is that they be finite.) The alternatives and the expected values of each choice are summarized in table 5.1.

How should we approach the question of rationality in a choice of this sort? From a practical or *strategic* point of view, Pascal's reasoning is rational.[7] Indeed, it has of late acquired a new sheen, in virtue of a widely trumpeted finding that churchgoers live longer and healthier lives (Strawbridge, Cohen, Shema, et al. 1997). The beauty of this new, pleasantly postmodern version is that it doesn't actually matter at all what you believe. For the tangible benefits of faith are not those promised by this or that sect in a life hereafter, but solidly secular benefits in this one.[8]

:: Table 5.1. Pascal's Wager

Life choice	*God exists* Probability $p_1 = 0.000000001$	*God does not exist* Probability $p_2 = 1 - p_1 = 0.999999999$	*Expected value* $\sum (d_i \bullet p_i)$
Ascetic faith	*Heaven* $d_1 = \infty$	*Futile asceticism* $d_2 = -1000$	$d_1 \bullet p_1 + d_2 \bullet p_2 = \infty$
Atheism	*Hell* $d_3 = -\infty$	*Life of Pleasure* $d_4 = 1000$	$d_3 \bullet p_1 + d_4 \bullet p_2 = -\infty$

Assume an extremely low probability p that God exists; assume also that a life of pleasure is worth 1000 units, while a life of self-denial or asceticism is worth minus-1000 units. The rightmost cells give the expected utility of each life choice, on the basis of the infinite value or disvalue of eternity respectively in heaven or hell.

Still, from an epistemological point of view, it is precisely this adoption of a strategic perspective in matters of belief that is irrational. ("Faith," quipped Mark Twain, "is believing what we know ain't true.") In effect, what enters into Pascal's strategic calculation is not *truth* but *probability* and *belief*. As if to compensate, belief actually figures in the equation twice over, playing two different roles. In the role of absolute faith, it constitutes the *act* we are enjoined to choose. Indeed, Pascal's originality lies in assimilating *what is rational to believe* to *what is rational to do*. But belief figures also in the calculation itself, in the form of *subjective probability*, quantified on a scale from 0 to 1. Call the latter "subjective belief."

Belief in the sense of what one does or doesn't do is what I have called "absolute belief": it doesn't admit of degrees. Subjective belief does admit of degrees. What is the connection between these two notions of belief? It might seem that an absolute belief that p is equivalent to a subjective belief that p that assigns to p a probability close to 1. But this supposition runs into a well-known puzzle known as the *lottery paradox*. Suppose that we fix the threshold of absolute belief at 0.999999. This amounts to saying that I believe (in the absolute sense) everything has no more than one chance in a million of being false. But now imagine a lottery of one million and one tickets, of which there is a single winner. Because each ticket has less than one chance in a million of winning, I have reason to believe (in the absolute sense) of each ticket that this one is a losing ticket. But all propositions that are the objects of absolute belief can figure in a deductive argument. In good logic, all conjunctions of true propositions are true. It follows, since ticket 1 will not win, and ticket 2 will not win, and so on for each ticket, that none will win. But since I also know that there *is* a winning ticket, I can deduce the contradiction: *All the tickets are losing tickets and there is one winning ticket.*

It would be fastidious to deck out all our assertions with a probability coefficient, and so, disregarding the risk of running into the lottery paradox, we treat most of our beliefs as absolute. We do not even require that the reasons on which these beliefs rest justify a probability value as close to 1 as in the lottery example. There is a disconnect, then, between the kind of absolute belief that figures in our everyday inferences and the Bayesian notion of belief that figures in rational decision theory.

From the biological point of view, as from the psychological point of view, believing what is true is not necessarily always the best strategy. This conclusion becomes more apparent when we consider an additional factor, as recently advocated by Christopher Stephens (2001). What determines the strategic rationality of a belief, Stephens tells us, is not just its truth, nor even its probability, but also (and sometimes most of all) its *importance*. His definition of this notion of importance amounts to saying that the importance of a proposition p is measured by *the difference the truth of p would make between the expected values of two actions*. Take a proposition like *there will be glazed chestnuts at the candy store*. Whatever its probability, its importance to me is low. Its truth or falsity will have little influence on the course of my life, no matter what I do. By contrast, the outcomes of Pascal's choices differ drastically, depending on whether or not God exists.

Stephens concludes that certain beliefs can be both *false* and *advantageous*, in that they increase the fitness of the organism. (A central example is the application of the "better safe than sorry" principle: better always falsely to believe that a predator lurks than to be right just nine times out of ten in the belief that none does.) In the long term, however, the rules of thought that will be selected are those that lead *in general* to justified beliefs.

It is instructive to compare this result with that of Sober on the evolutionarily stable strategies of truth-telling and lying. On the one hand, Sober predicts a point of equilibrium far from the two extremes. Individual instances of truth-telling or lying will be beneficial or harmful more or less at random: only at the level of the *population as a whole* will there be a point of equilibrium. Stephens, on the other hand, is interested in the strategy of *the individual* in regard to believing and disbelieving. For an individual, it is unwise to choose occasions of self-deception at random. One moral of this contrast between the individual and the population as a whole is that rationality at the social or collective level does not map in any obvious way onto individual rationality. Another is that the rationality of adopting a given strategy in general does not necessarily imply that the same strategy is always rational in any given case.

In a famous article, William Clifford stressed that even from a pragmatic point of view, truth is all-important. We have a duty to verify conscientiously all our beliefs, or else we are in danger of corrupting the very faculty of belief (Clifford 1886). From the point of view of epistemic rationality, however, we have already thrown in the towel by consenting to conduct the argument on such pragmatic grounds. The reasons Clifford advances invoke the possible *consequences* of error, and so they belong to the same category as the reasons that led Pascal to advocate bad faith. Whether bad faith is a good strategy or not becomes a purely empirical and practical question. But if truth is a norm that conforms to Ogien's criterion, it can't be evaluated in terms of its consequences. Consequences are variable, and their worth admits of degrees. On the strictly epistemic level, the success of a belief depends only on its truth. It does not admit of degrees and is relative neither to the interests nor to the circumstances of the subject.

But how, if not by recourse to pragmatism, do we justify a purely epistemic norm? According to the criterion proposed above, a norm is revealed as such by the criticisms to which its infraction is subject.

The problem is that, as soon as I open my mouth, I expose myself to criticisms of many kinds: for having spoken too loud; for grammatical errors; for hurting someone's feelings; for mispronouncing a word; for being immodest or repetitious. I could also be accused—and here we get close to the criterion of truth, but we are not quite there yet—of lacking sincerity or of risking my reputation on a doubtful proposition. But insofar as my assertion expresses a belief, only one criticism captures precisely the failure of this belief as such: that of not being true.

Now if I am to be sure that this is indeed the basis of the criticism to which I have been subjected, my objector must explicitly give the reason for her objection. The distinction between the two notions of belief, then, requires here again a linguistic capacity adequate for specifying that the objection mounted against my assertion is not any of those listed above, but specifically that it is *false*.

5.3 Why Other Animals Don't Reason Like Aristotle

An old debate still rages over the question: do other animals think?[9] No doubt, they think, common sense tells us, but *not like us*. (If a lion could speak, Wittgenstein said, we wouldn't understand him. No wonder Wittgenstein passed for a lion of philosophy.) One important difference between us and real lions is that a lion, however bold, will never risk an *assertion*. That's the larger part of the reason a lion can't be held responsible for telling the truth. While we may grant so-called lower animals certain powers of representation (*detection*, a primitive form of belief, and *tropisms*, primitive sorts of desires), we can best conceive of their behavior as resulting from the dynamic interaction of these representations, weighted by factors of *intensity*. By contrast, our own deliberations often take the form of the "practical syllogism," which dates back to Aristotle: "Every man should take a walk. I am a man. I should take a walk at once."[10] Intensity doesn't come into that sort of reasoning; the beliefs involved are absolute beliefs. Such reasoning cannot accommodate the more nuanced representation of thought processes implicated in real-life deliberation: "A walk would no doubt do me good, but swimming is even better for the cardiovascular system. And then, there are always risks, weak as they may be, and besides, there are so many other things I urgently have to get done...." Such a train of thought is doubtless beyond the reach of a lion. Nevertheless, it is probably a more faithful representation of the dynamic process of deliberation in animals, and that of humans as well when they deliberate nonverbally.

Though Aristotle himself attributes the practical syllogism to animals as well as to rational beings, the Bayesian calculus is better suited for representing what happens in beings capable of purposeful behavior when they weigh desires of varying intensity. This applies not only to animals but also to people when they exercise a physical skill that is not mediated by language, such as riding a bicycle or throwing a ball.[11]

Categorical belief does not exist in the Bayesian calculus. All "decisions" draw on a complex of forces. But we have no independent means of measuring the relevant parameters. This is why we cannot seriously tax an animal with irrationality. This brings us back to the evolutionary path, sketched over the course of the previous chapters, that led from simple reactivity or detection to planning and deliberate thought. Only at the end of this path, when we are dealing with a human being capable of full intentionality, can the grounds for a given decision be articulated explicitly enough to justify criticizing an agent for an irrational act or belief.

For those who are fond of theological analogies, we can discern here an equivalent, or perhaps even the origin, of the doctrine of free will. Those who lack free will enjoy the innocence of beasts; it makes no sense to blame a nonhuman animal. Yet the innocence such an agent enjoys is not the same as the innocence of an agent who freely chose the good while having the capacity to earn blame by choosing the bad.

5.4 Some Manifestations of Human Irrationality

If human beings attain their privileged status as rational animals by virtue of their capacity for irrationality, must we conclude that each instance of irrationality is a unique failure, or might we be *systematically* disposed to holding false beliefs and making invalid inferences? Psychologists have uncovered numerous phenomena that lend support to the latter hypothesis.

This has given rise to what has been called *rationality wars*. On one side, there are those for whom examples of a widespread propensity to make certain types of mistakes demonstrate that our intellectual tools are fundamentally defective. We are, they conclude, systematically liable to think and behave irrationally. Others see in these errors nothing worse than efficient shortcuts honed in our ancestors by natural selection and making up in speed what they lack in precision.[12] I will draw on a very small sample of examples to give some sense of the kind of errors at issue before moving on to consider what these cases can teach us about the evolution of rationality.

Consider first the all-too-familiar phenomenon of *superstition*. At once consoling and destructive to ordinary mortals, superstition is a source of inexhaustible profit for charlatans. You have a fever?

Come see me for an "alternative" treatment. The course followed by most fevers is this: they get worse for a while, then stabilize, and finally go away by themselves. More often than not, therefore, you will soon notice an improvement after my "treatment." Overflowing with gratitude, you will trumpet my wonderful methods wherever you go.

Superstition is, in fact, nothing but a side effect of a perfectly reasonable learning mechanism. Skinnerian learning, as discussed in chapters 2 and 3, rests on the propensity of an organism to increase the frequency of any behavior that is followed by a "reward." This propensity may be easier or harder to activate. The former case, known as an "error of the first type," consists in treating as significant what is in reality only a random statistical difference.[13] The converse, an "error of the second type," is erroneously to attribute to chance what is actually the effect of some real cause. The charlatan takes advantage of the first type of error, to which we are particularly susceptible. The problem is that if we are too keen to eliminate it, we risk the opposite error. There is no magic equilibrium point that guarantees an ideal result in all cases. If natural selection has honed our inductive faculty, we can expect this "hardness" parameter to be set a bit differently in different domains of learning. Indeed, there is evidence to show that this is the case (Cowie 1999).

Many of the most remarkable instances of systematic error discussed by psychologists arise from the difficulty of estimating risks involved in practical decisions. Risk induces fear. That is natural enough, since an organism reduces its chances of procreation, on average, in proportion to the risks it runs. Should we then expect to feel fear in direct proportion to objective danger?

For two reasons, the answer is no. First, the relation between the assessment of danger and the emotion it provokes (or for that matter the action that it motivates) is by no means linear. It seems reasonable to ignore a risk up to a certain level. Beyond a higher level, it seems equally reasonable to avoid *taking* the risk altogether. We allow ourselves to be guided by the intensity of our emotional reaction only in the band between these two levels. Only there, we might say, do we normally take risk seriously. We usually cross the road without being in the least afraid. As the traffic density increases, we begin to weigh the risks, and the psychic tension rises. At a higher level, we wait prudently for the protection of a red light, though even with this protection we are never guaranteed absolute security. In sum, we classify the strategies that we judge to be reasonable into discrete levels. The transitions between levels are something like phase changes in physics, such as freezing or evaporation. Below freezing, or above boiling, water behaves utterly differently.

The second reason that we shouldn't expect fear to be proportional to risk is of an entirely different order. It is due to the difference between the EEA and our own environment. This difference doubtless explains many of the systematic errors in reasoning that psychologists have imputed to us. Our ancestors acquired specific predispositions to experience fear in circumstances likely to afford the most frequent dangers. But our own circumstances differ from theirs. If we think of the appropriateness of fear as belonging to the domain of "axiological rationality," we can think of inappropriate fear as analogous to an instinctive reflex that has become useless in new circumstances. Hares flee in zigzags. This is an excellent tactic for escaping a predator, but it proves catastrophic when the hare is on a highway, and what threatens it is a motorist seeking *not* to hit it.

For ourselves no less than for the hare, our atavistic reactions are often still useful. A menacing-looking stranger; a snake; a large animal; and by extension, a big machine that bears down on me: all this rightly provokes fear. Similarly, the disgusting taste of many poisons keeps us from ingesting them. But we should also expect important differences between the reactions that are programmed or facilitated in the EEA and what is sensible to fear in our present ecological niche. Objectively speaking, cars are ten times more dangerous than airplanes. However, even if we could precisely calibrate and measure our fears, we would rarely find someone who fears cars ten times more than airplanes.

Psychological research has laid bare numerous systematic distortions of our reasoning or our emotional reactions.[14] Common attitudes toward risk constitute but one example and seem perverse in subtle ways.[15] One particularly well-studied range of cases concerns estimates of risk on the basis of numerical data. Here is an example.

> The Base Rate
> Suppose you are given a test used to diagnose a certain type of cancer. This type of cancer is known to affect 0.001 percent of the population. We also know that the results of the test are 98 percent reliable. Your test results come back positive. What should you conclude about the probability that you have this type of cancer?

Many people infer from the fact that the test is rated 98 percent reliable that the probability is very high—certainly over 50 percent. This seems to follow from the fact that the test is said to be 98 percent reliable. Actually, however, the probability that you have cancer in this scenario is under one half of one percent. The common mistake stems from a failure to take account of the *base rate* of cancer cases—that is to say, the actual frequency of cases of this type of cancer in the population as a whole. Since this cancer is actually very rare, the risk indicated by the exam result is much lower.

In order to see why, let us look at the problem in a different light. To keep things simple, assume that the reliability of the test is the same for both positive and negative results. (This wasn't specified in the original data, and it might not be the case in general.) That says that for every 100 people who have cancer, 98 are detected and 2 are missed, while for every 100 people without cancer, the test is correct 98 times but twice erroneously turns in a false positive. Now let's imagine that 1,000,000 people take the test. According to the base rate, we can expect that about 0.01 percent or about 100 of that number people have contracted the disease, while 999,900 have not. We know that 98 percent of cases should be detected by the test. The test will also give us 2 percent false positives, that is, 19,998 out of the 999,900 healthy cases. Of the 1,000,000 tests taken, therefore, we can expect that 20,096 will show positive results, of which 98 are correct while 19,998 are false positives.

It follows that the chances of a subject whose test result is positive actually suffering from this form of cancer are a mere 98/20,096—which is 0.00487 or less than 0.5 percent.

In the face of such findings, Gerd Gigerenzer and his colleagues (1999, 2001) have taken it upon themselves to champion the rationality of the human race. According to them, the difficulties we encounter when trying to execute the calculations specified by decision theory in this sort of case result from two exonerating factors.

First, we can't generally assume that the subject of the test was randomly picked. You don't take a blood test unless you worry that you might be sick. In this case, the base rate would not necessarily be relevant. It might seem quite reasonable to ignore it.

Second, Gigerenzer suggests that we should pay attention to the way the problem is presented. The way we formulate a problem doesn't logically affect the solution, but it can make a big difference to the ease of finding it. The trick to working out the solution often lies in the discovery of an appropriate algorithm, and an algorithm presupposes a certain mode of representation. Unless the notation is appropriate, the algorithm can't be applied (Marr 1982). Simple addition, for example, is greatly facilitated by an elegant algorithm:

- Line up the numbers to be added, justifying on the rightmost of each;
- Add the rightmost digits;
- Write down the result if it is expressible in a single digit;
- Otherwise, write down the rightmost digit of the resulting number, and retain all the digits besides the rightmost one;
- Move left one digit and repeat, adding the digits retained if any;
- Repeat, moving leftwards, until there are no more digits.

But this works only if the numbers are represented in a suitable notation. Arabic numerals work fine, but if we try to plug in Roman numerals, the algorithm is useless. This applies just as well at the level of psychological mechanisms. In chapter 3, I described the difficulties faced by most people in trying to solve Wason's puzzle, and we saw that the problem largely disappears once the problem is presented in such a way as to activate the "cheater-detection algorithm." To the extent that the structure of our cognitive faculties is modular, certain representations will be more accessible than others. Where a module has been honed by natural selection to perform a specific task in a given domain, an approximate answer to the right kind of problem will be computed quickly. But if the terms in which the problem is set are not those on which the module has been "trained," the same problem may seem as intractable as addition for Roman numerals. In the case of the diagnostic test just mentioned, for example, the data were originally formulated in terms of probabilities. The equations involved in the calculus of probabilities are complicated and difficult to manipulate, requiring the unfamiliar language of conditional and a priori probabilities. By contrast, talk of frequencies or ratios appeal to more familiar intuitions: when made explicit in terms of such ratios, the reason the cancer risk remains low despite the positive test becomes easier to grasp.

I have repeatedly stressed that rationality is not equivalent to success. An equally important difference holds between the rationality of a global strategy and rationality as it applies to a particular case. The "strategies" attributed to natural selection are conceived in the most general terms, on the basis of past experiences and without precise information about the future circumstances under which they may be implemented. By contrast, individual rational agents face particular problems in possibly unique circumstances. On the basis of this contrast, even someone who recognizes that a general requirement of truth in beliefs is implied by the necessity of getting around in the world might hold that the strategic attitude to belief can be reasonable in specific cases. Hence what plausibility Pascal's wager may have: we can grant that true beliefs are a good idea in general but insist that holding a false belief might sometimes be most likely to achieve our ends.[16]

The modules acquired in the EEA may become maladaptive in a changed environment. When that happens, we may have to resort to a system of representation that cuts across different domains. As suggested by the work of Carruthers and others mentioned above (in chapter 3), language with its mathematical extensions is just such a system. The tools of explicit reasoning are slow but powerful. Their strong point is their capacity to negotiate among heterogeneous interactions, whether among different sensory modules, between sensory modules and other cognitive modules, or among the latter modules themselves. When pitted against the most efficient modules that have evolved to handle certain types of problems, explicit reasoning falls short in both elegance and speed, just as the integral calculus applied to a steam engine leaves much to be desired compared to the elegant efficiency of Watt's governor. But when the different cognitive domains need to talk to one another, the explicit mode is at times indispensable.

The cases referred to so far illustrate three themes that are key to the present work: the contextual relativity of rationality; the modularity of our cognitive faculties; and the bifurcation between the "intuitive" and "explicit" modes of reasoning, of which the first are based on analogical systems of representation that dynamically determine our behavior, while the second are grounded in digital representations. Once we take account of the relativity of rationality assessments that follows from these multiple perspectives, the quarrel over rationality seems pointless. There are other cases, however, that can't be dismissed so easily.

The Normative Transitivity of Preference

In chapter 4, I called attention to the difficulties that we encounter when we try to transpose the notion of rational preference, with its generally accepted requirement of transitivity, from the individual to the group. I now want to suggest that the norm of transitivity of preference is problematic even for the single individual.

Let us assume that we can quantify the value of different outcomes. The values thus obtained could then figure in a trivial arithmetical equation that captures the principle of transitivity (T):

$$(T)(A > B) \,\&\, (B > C) \rightarrow (A > C)$$

Despite its plausibility, however, this principle seems violated by a simple children's game. Rock, Paper, Scissors is based on the relation of dominance, plausibly represented by the "greater than" symbol ">" in (1). This relation of dominance is circular: paper smothers rock, rock crushes scissors, and scissors cut paper. Why should our preferences never be determined by mechanisms structured in this way?

If I am asked whether I prefer Braque or Picasso, I will choose Braque. Now add Matisse to the list. Suppose I say I prefer Picasso to Matisse, but also that now I prefer Matisse to Braque. In all, then, my preference ranking is:

$$B > P > M > B.$$

Must we say that in the meantime I changed my mind? In any case, like Sidney Morgenbesser in the anecdote cited in chapter 4, I have violated the condition of independence of irrelevant alternatives. But here is how we can make sense of it. The introduction of the Matisse option, irrelevant to the first list, caused me to think of some new criterion. In the first instance, I thought of both painters as Cubists. Among Cubists I prefer Braque. But now that we have broadened the field, I realize that I prefer Matisse to Braque and also prefer Picasso to Matisse. Nevertheless, when considering just Picasso and Braque, the original criteria recover their salience, and I maintain my initial preference.

Several philosophers have worked out various similar scenarios in which transitivity seems to have been violated without thereby making the choice indefensible.[17] Michael Philips (1989) further argued that if these examples compel us to give up on the transitivity condition, then the entire theory of rational decision is undermined. Without following him quite that far, it must be granted that it is difficult to guarantee in advance that no new option could trigger a change in an agent's criteria of choice. And unless we can vouch for the continued independence of irrelevant alternatives, the norm of transitivity can hardly be of much use.

This suggestion will no doubt meet with great resistance. Some will insist that transitivity is not a descriptive natural law that can be disconfirmed by simple counterexamples. It is a *norm* of rationality. To infringe it is not impossible, but it is irrational. Any impression to the contrary results from equivocations that invalidate the alleged counterexamples.

Three Levels of Normativity

The objection just formulated rests on a simple opposition between the normative and the descriptive. Now recall the findings, cited in chapter 4, that showed the effectiveness of sanctions in enforcing norms. The content of the norms in question undoubtedly has a world-to-mind direction of fit; but the enforcement of the norms is constituted by natural facts—the existence of sanctions—the *description* of which has a mind-to-world direction of fit. It might well be, therefore, that other norms of rationality that interest us also owe their existence and force to social or psychological *facts* that are not immune to counterexamples.

To support this naturalistic speculation, I propose to distinguish three levels of normativity. For each, I will provide a typical example. Remember that I stipulated that normativity, in the sense that interests me, is tied to the possibility of legitimate criticism. Furthermore, every norm analytically corresponds to a specific criterion of success. Without contravening Ogien's rule that true norms imply "oughts" that don't admit of degrees, norms of success can be more or less stringent. Curiously, we shall see that the more stringent the norm, the harder it is to understand the criticism that can be raised in its name.

Consider first the principle of *noncontradiction*. If two propositions p and q are incompatible, it seems natural to criticize whoever believes both "that p" and "that q" for being incoherent, and thus irrational. But suppose someone is accused of believing an explicit contradiction, p *& not-p*. This charge now seems to make as much trouble for the critic as for his or her target. For what proof could the critic possibly have that such a formulation truly represents someone's beliefs? It seems that the very attribution of a direct contradiction such as p *& not-p* is itself incoherent. In support of this contention, we could follow Quine and Davidson in invoking a "principle of charity" that enjoins us from attributing directly contradictory beliefs.[18] But actually there's no great charity in giving away what you can't hold onto. What's wrong with attributing an explicitly self-contradictory belief is a little like what Hume complained of in claims about miraculous events. Though a miracle is not logically impossible, it may still never be reasonable to believe in one. For the alternative hypothesis will always be more plausible, to the effect that you are either mistaken about the facts or ignorant of the applicable laws of nature. Similarly, to the attribution of a self-contradictory belief, the alternative hypothesis will always be more plausible, that you have misinterpreted your interlocutor.

The upshot is that the law of noncontradiction seems to be a very curious norm: one so strong that it is *impossible to infringe it*. Should we infer that it is a *law of nature* that one can't believe (p & *not-p*)? This would contradict the dogma that insists on the radical opposition between a natural law and a normative rule. Here it seems that the two actually merge: it is because we *cannot* believe in a contradiction that we *ought not* hold a belief that implies one.

The case is somewhat different with a second, more questionable, category of norms of rationality. This category comprises the theorems and most of the principles of inference of mathematics and logic, including the calculus of probability. We often get those wrong; yet in most cases we can demonstrate their truth. An excellent illustration is provided by a puzzle long familiar to decision theorists and now widely known as "the Monty Hall problem" after the American television host who exploited it. Here is a simple variant of that puzzle:

> Three cards, two kings and one ace, lie face down on the table. I ask you to place a finger on one of these cards. I then turn over one of the others, revealing a king. We know that

the two cards remaining are a king and an ace; your finger rests on one or the other. You are now given the chance to bet that you are pointing to the ace. To give yourself the best chance of winning, should you: (a) move your finger to the other card; (b) stay on the same card; or (c) do either one indifferently?

Many people choose (c). They reason that an ace and a king remain, so you have an even chance either way. But that is the wrong answer. To convince yourself of that, think of what would happen if you played the game over and over. There are two kings and one ace at the start of each game. So the chances of your having your finger on a king at the start are two in three. It follows that if you wish to point to the ace, you will succeed twice as often by switching.

This argument is perfectly conclusive. But now consider a third sort of case, made famous by Robert Nozick: Newcomb's problem:

> A being in whose power to predict your choices correctly you have great confidence is going to predict your choice in the following situation. There are two boxes, B1 and B2. Box B1 contains $1,000; box B2 contains either $1,000,000 ($M) or nothing. You have a choice between two actions: (1) taking what is in both boxes; (2) taking only what is in the second box. Furthermore, you know, and the being knows you know, and so on, that if the being predicts you will take what is in both boxes, he does not put the $M in the second box; if the being predicts you will take only what is in the second box he does put the $M in the second box. First the being makes his prediction; then he puts the $M in the second box or not, according to his prediction; then you make your choice. (Nozick 1993, 41)

A straightforward Bayesian argument recommends maximizing expected utility. Providing the oracle inspires sufficient confidence, this clearly implies that we should take only one box. If, for example, I attribute a 0.9 success rate to the oracle's predictive powers, the expected desirability of taking just one box would be $(0.9 \times 1,000,000) = 900,000$, as against $(0.1 \times 1,000,000 + 1,000) = 101,000$ for taking both. But a "dominance" argument for the opposite conclusion seems equally compelling. When I make my decision, either the money is already there, or it is not. Nothing could change this now. Regardless of the oracle's prediction, then, the choice to take both boxes dominates: whether or not the million is in the first box, I will get $1000 more if I take both. Subjects divide about equally on the answer they favor, but in each camp most people tend to be thoroughly convinced of the other choice's irrationality. No more fundamental principle seems to be available to decide between the two arguments.

In the 1993 text from which I quoted above, however, Nozick indicates that the intuitions upon which each position is based can be manipulated by modifying the amounts specified in the problem data. If the oracle only placed a single cent in the second box, then even the most committed advocates of the dominance argument are unlikely to risk taking both boxes. Conversely, if the amount promised for the first box is relatively insignificant, Bayesians will be far less likely to stick to their principles. When two well-established principles conflict, most people will make their decision on the basis of vague intuitions, of feelings of conviction that resist analysis. In contrast with the strongly constraining principles that dictate the correct solution of the Monty Hall problem, no principle seems to be ultimately compelling in the Newcomb case. The principles involved are *optional*. If anything is decisive, it is something like emotional intuition, and that varies from one person to another.

5.5 Temporality

The dominance of feelings in the absence of conclusive rational principles is even more striking when we consider our attitudes concerning the future, the past, and duration. In these areas, there is no lack of opinions about what is rational and what is not, but most convictions are built on the shifting sands of our emotions.

Philosophers define *weakness of the will*, or "akrasia," as the choice of agents to act against their considered judgment of the best reasons, all things considered. With the exception of Oscar Wilde who claimed he could resist everything except temptation, most of us take it as axiomatic that it is rational to struggle against weakness of the will. But why? One standard response is that those desires that we give in to as a result of weakness of the will thwart our long-term desires. The superiority of the long-term perspective, however, is incontestable only from the long-term perspective. Robert Nozick, in an uncharacteristically egalitarian mood, observes that each moment should count equally when the final reckoning comes. On that assumption, we can determine in principle when resisting temptation is better overall. It will be so whenever giving in to momentary temptation will garner, in the long term, a sum of regret—weighted by the time it will last—outweighing the total moments of pleasure.[19] But how are such sums computed? It seems logical to expect that the retrospective value of all moments of experience would be equal to the value of each moment, weighted by its duration. But Daniel Kahneman has shown that this doesn't fit the facts. When we evaluate a series of past moments, it seems we care disproportionately about just two moments: the one that had the most extreme value, whether positive or negative, and especially the moment that came last

(Kahneman 2000). This finding suffices to invalidate just about any method you might devise for measuring your happiness. For the result will vary with the time you choose to compute it. And at the moment when you are in the grip of temptation, why should you be concerned with those other moments?

For natural selection, an individual organism's survival matters only as long as it takes to transmit the information that is briefly entrusted to it. We can surmise that Kahneman's formula for evaluating the past could benefit our genes. Whoever would scrupulously subtract from the joys of motherhood the pain involved in childbirth, for example, might not choose to repeat the experience—to the detriment of some gene lineages. But from my point of view as an individual, I see no particularly compelling reason to adopt one attitude to the past in preference to any other.

The same goes for the future. One factor that predisposes us to akrasia is that we discount the future. Up to a point, this is a perfectly reasonable way to take account of uncertainty, of inflation, and of interest rates. But the rate of discount implied by the choices made by supposedly rational agents is spectacularly underdetermined by those factors. As George Ainslie has shown, humans discount the future at a hyperbolic rate (Ainslie 2001). Nonhuman animals do this too (Ainslie and Monterosso 2003). This implies that the order of preference of two future events can be reversed as we get closer in time. It is similar to what we can observe when we get close to a small building that is located in front of a skyscraper. At a certain point, the smaller building blocks out the larger one. Ainslie, like Nozick, views this as a kind of pathology of the will, caused by an imperfect mechanism for the detection of relative value. More recently, Martin Daly and Margo Wilson have demonstrated that this pathology reaches even farther than Ainslie anticipated. The rate of discount applied by men to a monetary offer increases sharply when they have just viewed photos of attractive women. This increase is unaffected by the fact that there is not the least likelihood of the subjects' meeting the models. Perhaps unsurprisingly, no analogous phenomenon affects females (Daly and Wilson 2003).

Once again this result tends to confirm the existence of specialized mental modules. In the EEA, it was no doubt useful to act in accordance with the principle that a bird in the hand is worth two in the bush. In the situation just mentioned, the module that applies that principle is relatively impermeable to the information, however evident to the rational mind, that the sight of a photograph alone affords no opportunity to procreate. Such an explanation invokes the vestigial teleology of evolution, but that biological end obviously doesn't justify the attitude in question. On the contrary, in light of a broader consideration of rationality, the observed increase in the discount rate is entirely irrational. Subjects who managed to override that automatism probably resorted to explicit calculations in light of realistic considerations about the value of money through time.

The specialized abilities in the toolbox bequeathed in our brains by natural selection were those that our ancestors needed to face specific concrete problems. This suggests a reformulation of Plantinga's challenge referred to in chapter 1. How could it be that, in spite of the modular structure of our mental faculties, we were able to discover the general rules of logic and epistemology that allow abstract, domain-independent reasoning?

In chapter 3, I adopted Peter Carruthers's hypothesis that language is what enables us to navigate between the different modules (Carruthers 2002a; 2002b). Since mathematical algorithms make use of explicit linguistic expressions, this hypothesis affords a solution to Plantinga's challenge. Once language enabled us to refine genuinely topic-neutral inferential strategies, we no longer had reason to fear that legitimate inferences might be confined to those practical contexts that occasioned the development of specialized cognitive modules.

That conclusion does not, however, justify unbridled optimism about our powers of rational inference. We tend to think of mathematics and logic as providing an ideal of rationality that promises eventual solutions to all intelligible questions, if only we remain sufficiently steadfast in their pursuit. As Christopher Cherniak (1986) has shown, however, the idea of a perfect rationality is utopian, implying, among other absurdities, that all mathematical truths should be immediately transparent to the rational mind.

For that matter, purely logical rationality does not suffice to resolve even the simplest dilemmas. Recall, for example, that a wager is fair if and only if the expected value of placing it is identical to that of not betting at all. So how can expected value help us decide whether to bet or not? Once again, it seems that only *emotional attitude* differentiates those who prefer adventurous risk from those who choose prudently to sit it out.

The examples of nontransitivity discussed above yield a similar moral. We saw that apparent infringements of transitivity might be explained by the activation of different sets of criteria. But when we observe the behavior of others, we have access only to the outcome of their choices, not to the criteria upon which those choices are based. Thus we can never be sure of being justified in criticizing another agent's choice. A principle of charity intervenes, though here we are not tempted to suppose that transitivity is a law of nature. On the contrary, since our choices result from the complex interaction of highly diverse factors, we should not expect that they will fit a simple mathematical model. By contrast, when I find myself violating transitivity (or going in for the sort of apparent non sequitur illustrated by Morgenbesser's quip), I should take it as a call to delve more carefully into my own motivations.

5.6 Abstraction and the Frame Problem

Any representation is bound to omit the greater part of the represented item's characteristics. If we try to make up a counterexample to this claim, we'll soon find ourselves devising something like Lewis Carroll's one-mile-to-the-mile map (Carroll [1893] 1982, 717). But even a scale of one mile to the micron would not suffice to display every feature of the mapped terrain. All the representations we actually use are inevitably shortcuts, approximate but efficient, that we have developed over the course of evolution.

Two types of principle govern the process of sorting features to be represented in any abstraction from those that get left out. First, the structures and capabilities of the brain itself will favor certain categories of information: the colors that our trichromatic visual system constructs for us, for example, or the range of details picked out by our cognitive modules as relevant to our likely practical goals. Second, and too easily forgotten, there is a repertoire of emotional factors that favor schematic simplifications in the way we perceive situations. These are indispensable in practice, but they also give rise to prejudice, distortion, and self-deception.

When we consciously envisage the selection of properties to be included in a representation, we run into a problem that arises neither for natural selection, nor for the modules that natural selection has refined. This is a puzzle known as the *frame problem*. This problem first arose for the theory of knowledge representations in artificial intelligence.[20] It results from the fact that any action has an indefinitely large number of consequences. Some are intended; others are mere side effects of which we are generally quite unaware. Every time I take a step, millions of air molecules and particles of dust are displaced. Normally, this is of no significance. However, under certain laboratory conditions—in the work of nanotechnology, for instance—ignoring these events could be as dangerous as lighting a match in a tank filled with gasoline vapor. How could we program a robot to pick out just the relevant consequences of any action?

Imagine that our robot already has an encyclopedic knowledge of all subjects. In managing this labyrinth of information, it must focus on those consequences that are (or may become) significant and avoid wasting time on the contemplation of irrelevant ones. It is important to see that this problem goes beyond the classic problem of induction. The latter concerns the question of the justification of inferences. In the frame problem, however, the question is not how to justify a given conclusion, but *how we are to know whether a conclusion is relevant before we bother to draw it*. We need to ignore the greater part of the immense field of possible inferences. But such a demand is paradoxical: for it seems that we would need to examine everything in order to know what we would be entitled to ignore.

5.7 Emotions and Rationality: The Axiological Empire

The frame problem does not exist for sensory modules, because for the most part, these are isolated from one another. My ear, as such, has no decision to make about whether to ignore gustatory information, because no such information ever reaches it. For the same reason, if there are cognitive modules, the frame problem is no threat to them either. By definition, each module fulfils its function on the basis of a limited domain of information. However, the frame problem is exacerbated, if not created, by the proliferation of possibilities of representation made possible by language. Why then we are not paralyzed by even the simplest choice?

I have no complete answer to this intractable problem. But a partial answer, I suggest, lies in our emotions. Emotions constrain and focus our attention and thus help us evade the information gridlock threatened by the frame problem. By temporarily restricting the scope of our attention, emotions allow us to ignore much of the information at our disposal. (The emotional plea, "Don't confuse me with the facts!" is not always wholly absurd.) Our emotions mimic the modularity of our sensory channels and restrict the information that we take account of. Such a restriction is variable, usually temporary, and sometimes, alas, deplorably counterproductive. Feeling angry at my friend, I will see only her faults. Later, having made up with her, I will forget all of her failings and set myself up for future disappointments and resentments. Or again, while in the grip of fear, I will be unable calmly to assess the real dangers of my situation. Often my decision will be in no way justified by statistical facts of which I'm well aware. That is why emotion is often viewed as the cause of irrational behavior. But the charge of irrationality must be mitigated with the reminder that any useful function will inescapably have undesirable side effects in some circumstances. We saw this general truth at work in the case of superstition, in which the side effects in question were necessary consequences of the basic mechanisms of belief acquisition. Equally vast are the effects of this general truth in the domain of practical reason. We often complain of the blinkered, emotionally driven behavior of others; but in fact, the mechanism that produces such blinkering is a condition of rationality itself.[21]

To be in the grip of an emotion is in some sense to apprehend a value.[22] But is this apprehension a *perception* of some external reality or a *projection* of emotion? This is an old question, raised by Plato in the *Euthyphro*. Euthyphro having defined piety as "what is loved of the gods," Socrates asks whether the gods love it because it is pious, or whether it is pious because the gods love it. Socrates' question can be applied to everything we hold good or beautiful: do we value and aspire to what is good or beautiful because it really is so, or are "beautiful" and "good" merely labels for whatever we value and aspire to? I will not try to answer this deep question here. I will confine myself to a more modest claim: that there is a domain of value, which I call the *axiological* domain, to which emotions are pertinent, and that the criteria of axiological rationality are not reducible to those of either strategic or epistemic rationality. Often, as we have seen, the criteria in question are counsels of wisdom that concern our attitudes to duration, the past, or the future: domains that appear to remain imponderable for classical decision theory.

Here are two specific examples of plausible principles of emotional rationality over time. The first, the *Philebus principle*, enjoins a proportionality between the anticipated pleasure and the pleasure that we get from anticipation. According to this principle, it is irrational to take great pleasure in the expectation of something that one knows will bring little or no pleasure when it comes. Conversely, it seems quite reasonable to dread the prospect of future suffering. The Philebus principle seems quite reasonable, even though it conflicts with a purely utilitarian, economic principle we might label the "get-what-you-can" principle. Get-what-you-can says that if you are not going to enjoy some future event, you might as well enjoy the prospect of it if you can. That way you'll have got *something* out of it. And as for the case of future suffering, there's no point in making it worse by adding the pain of anticipation. Which principle should we live by? Get-what-you-can has a sort of mad logic, but most of us couldn't just opt to disconnect our feelings of anticipation from our assessment of our future pleasure or suffering. Think how confusing ordinary practical reasoning would be: "I'm enjoying the idea of the pain and suffering that will result if I marry Peter, and I'm pained by the prospect of the joy it will be to marry Paul. So which should I do?"

Giving up the Philebus principle would deprive us of the guidance afforded by our present emotions respecting the likely consequences of our current choices. So the anti-utilitarian Philebus principle seems to have a practical justification after all.

The second example of a principle of purely emotional rationality is an *Aspectual principle*, which urges us to take account of the difference between activities that take place in a continuous interval of time and others that are appropriately seen as achievements, to be completed at a given point in time. This last principle recommends, for example, that we appreciate our pleasures as they take place in time, as distinct from the satisfaction of having enjoyed them (de Sousa 2000). It implies that such activities as viewing art or making love are best left to those who concentrate on enjoying them as they unfold, in contrast to those who are anxious in getting things done and checking off accomplishments.

While both the Philebus and the Aspectual principles could be recommended on the basis of persuasive reasons, I don't think either is reducible to any straighforward precept either of practical or of epistemic rationality. Both rest crucially on their emotional appeal. In this they illustrate the relative autonomy of the axiological domain.

But the axiological is more than merely autonomous. Our emotions are actually capable of playing the role of mediators or arbitrators in the sort of confrontation between the principles of epistemic and of strategic rationality with which the present chapter began. Only our emotional sensibility is up to the task of judging— with the aid, of course, of all manner of reasoning, evidence gathering, and reflection that our intellectual faculties afford— whether a given situation calls for the application of an epistemic principle or a strategic one. It was epistemic passion, for example, that pushed Clifford to object to the purely strategic attitude adopted by Pascal to the question of the existence of God. Adjudication between them cannot belong to either, on pain of letting one or the other be both judge and party. Adjudication properly belongs, I suggest, to axiological rationality, which thus plays the role of supreme arbiter with respect to the two traditional domains of rationality.

The critical role of emotions in the economy of rationality was already anticipated by Descartes, who wrote:

> The utility of all the passions consists just in this: that they strengthen certain thoughts and maintain them in the soul, which it is good for the soul to conserve and which might otherwise be easily effaced. In the same way, all the harm they can do consists in the fact that they fortify and maintain these thoughts longer than is needed, or that they fortify and conserve other thoughts on which it is not good to dwell.[23]

Translated into the perspective of modern cognitive science, the "utility" as well as the "harm" of the emotions are due to the modular organization of a complex emotional-cognitive system. Over the course of time, the elements of this system were cobbled together by natural selection, more or less independently, and more or less harmoniously, and they often remain relatively autonomous. This genealogy has bequeathed to us the capacity to transcend the limits of practical rationality laid down by natural selection, but it has also exposed us to conflicts of value. It also necessarily brings an extensive range of potential irrationalities, in both thought and behavior, and at both group and individual levels. Perhaps we should conclude that the logic of the processes that have brought about such an architecture of the mind is ineluctable. There would then be no escaping the deeply demoralizing conclusion that we do indeed live in the best of all possible worlds.

Notes

Chapter 1

1. At the time of writing, agents of Honda, Inc. are touring the world with a walking robot called Asimo. This robot, which was seventeen research years in the making, is endowed with a dignified gait with which it walks and climbs stairs. Notwithstanding this impressive achievement, however, Asimo would be quite incapable of beating a two-year-old child in a running race. This compares unfavorably with Deep Blue's performance in beating the world's greatest chess masters at their game.
2. For any item in a language, linguists distinguish types from tokens of that item. A token is a particular occurrence of a type: in the string *zzzz*, for example, there are four tokens of a single type.
3. This variety, and its significance for the analogical faculty that makes it possible for us to recognize them all as instances of the same letter, is beautifully illustrated and discussed by Doug Hofstadter and his colleagues in Hofstadter and McGraw (1995).
4. See Ridley (2000). He cites (on p. 83) a mutation rate for humans of about 3×10^{-8}. Ridley speculates that such a rate is not far below that which would lead to "mutational meltdown," entailing the disintegration of our species.
5. See Plantinga (1993). For critical discussion, see Fitelson and Sober (2001).
6. Indeed, in the Judeo-Christian tradition to which Plantinga belongs, the Tree of Knowledge is precisely what was forbidden to original Man and Woman. If it is objected that the prohibition applied only to the knowledge of good and evil, that presupposes that the deity held an austerely modular view of the different aspects of knowledge (we'll be looking at exactly what *modular* means in chapter 3.) But it is not clear that the "module" of evaluative knowledge can be so neatly sequestered from all other domains of knowledge.

Chapter 2

1. For a gallery of unlikely monsters, see the photographs brought back from the deepest ocean by the Norfanz expedition, available online at http://www.oceans.gov.au/norfanz/CreatureFeature.htm.
2. This view once enjoyed a certain vogue. It was defended in a number of little red books published by Routledge & Kegan Paul about half a century ago, such as Winch (1958) and Melden (1961).
3. A joke, due to Robert Gordon, sums this up quite well. Why do you so badly want a Mercedes? Because if I didn't, there would be no chance of my buying one!
4. Distinctions to which Austin ([1956] 1970) wittily adverted, and which come under detailed scrutiny in Bratman (1987).
5. Pierre Simon, Marquis de Laplace (1749–1827) published his *Exposition of the System of the World* in 1796. This work defined a classical version of the doctrine of determinism. Laplace's point of view is now no more defensible than that of Aristotle, described in the paragraphs that follow in the text. The first scientific discovery to shake Laplacian determinism was thermodynamics; then came chaos theory; but the blows that definitively sealed its fate, so to speak, were, first, quantum mechanics and, second, the possibility, which remains a live one, that stochastic phenomena might underlie all other apparently deterministic ones. See Bohm (1957) and Prigogine (1994).
6. For the classic exposition of the distinction between singular causal relations and explanatory statements, see Davidson ([1967] 1980).
7. See Kauffman (1995). The relation between Kauffman's proposal and the Aristotelian concept of immanently organizing essence has been explicitly recognized in Walsh (2002). In chapter 4, I shall sketch the basic principles of "cellular automata," which are models relatively close to those with which Kauffman's speculations are concerned.
8. This should not be confused with the claim that getting an orderly *looking* sequence of some kind is no less likely than getting some random-looking sequence. That is false, of course, since the number of orderly looking sequences of any given length is far smaller than the number of random-looking sequences of the same length.

9. See Lem (1974). I shall argue in chapter 5 that no degree of probability, however high, suffices to establish that some proposition is true or should be believed without qualification.
10. One can get an idea of the power of the combinatorial explosion involved in this sort of calculation by looking at something entirely different. The French writer Raymond Queneau has left us the collection *One Hundred Trillion Sonnets* (Queneau 1961). It consists in a mere ten pages. On each page are fourteen lines of a potential sonnet, featuring a single identical rhyme scheme. The individual sonnets of the collection are obtained by taking one line from any of the ten available ones, followed by a line from any of the next, and so forth. Playing with just the first two lines already yields 10^2, or one hundred different sonnets. Since there are fourteen lines, the recipe suffices to generate 10^{14}, or precisely one hundred trillion sonnets. You can see how this works and generate some of these sonnets at will at a site set up by Magnus Bodin at http://x42.com/active/queneau.html.
11. In the fourth *Meditation*, Descartes asserts that "there is considerable rashness in thinking myself capable of investigating the impenetrable purposes of God" (Descartes 1986, 39); that doesn't stop him from speculating just a few pages later, however, that "I cannot . . . deny that there may in some way be more perfection in the universe as a whole because some of its parts are not immune from error, while others are immune, than there would be if all the parts were exactly alike" (ibid., 42–43).
12. See Lewens (2002, 10–11) for a list of possible explanations for the imperfections of the biological works of the creator. Lewens cites George C. Williams as the source of this example. Two more obvious examples: the proximity of the tubes we use for respiration and swallowing puts us at risk of choking whenever we eat while breathing; and the receptors on our retina turn their backs on the light that it is their function to capture.
13. There are two principal varieties of teleology: goal or purpose and function. One can say of a tool that it has a function rather than a goal, but it was with the goal of serving such a function that the tool

was designed. A goal, then, will commonly be a certain state of affairs, while a function will more likely be identified with a specific means of achieving that state of affairs. A goal is typically particular: so an act can have a novel goal. By contrast, the function of an object or an organ is a property of that type of object or organ in general. It is true, of course, that an object can be used to serve a once-only function: a dictionary may serve as a doorstop; but the "function" assigned to it in those special circumstances will not displace or change its "normal" function. Moreover, a function is always served in the context of a whole; the thriving of the whole gives the function its point. If the car is smashed up, the accelerator no longer serves any purpose. A goal, by contrast, can be set by a single arbitrary intention, without any necessary connection with any system of value or utility. From the statistical point of view, which is that of evolution, the notion of a particular goal gets no grip. Yet we can sometimes speak of general goals pertaining to mechanisms that bring about certain results that are deemed "normal." In what follows, I will not further differentiate between teleology, goals, purposes, and functions.

14. Some recent scholarship has cast doubt on this assumption. See, for example Lennox (2001) and Winsor (2003). It isn't clear, however, how Aristotle's conception of the relation of actuality to potentiality could be compatible with the idea of evolving species. For that would seem to imply that in some cases the actual is not "prior" to the potential, as is explicitly required in Aristotle's texts. See for example *Metaphysics* IX–8 (Aristotle 1984, vol. 2).

15. Admittedly this fact continues to be stubbornly denied in certain circles. The Aristotelian conception is still current in the Vatican and popular with all those who oppose genetically modified products or artificial reproductive technology on the ground that these are "against nature." All these modern Aristotelians are one in spirit with the airline passenger in an old Gardner Rea cartoon who remonstrates with the flight attendant: "No thank you, I don't think Nature intended us to drink while flying."

16. This formulation is highly simplified. It is inspired by Proust (1997, 216–17). It purports to define only the notion of "direct proper function." Other notions such as "derived proper functions" of semantic elements of language can be defined in terms of this. This theory has been subjected to a great deal of debate and discussion, but the refinements need not concern us here. Many of the most important debates are conveniently collected in Allen, Bekoff, and Lauder (1998).
17. On the spread and persistence of religion, see Pascal Boyer (2001).
18. Given the still-heated recent discussions of this problem, this may seem a little short as well as vague. Denis Walsh and André Ariew (1996) have offered an irenic solution that rests on the requirement that function attributions be relativized to certain interests and to a "selective regime" effective either now or in the past. This is an attractive proposal but has the drawback of diluting the objective character of the function attributions made strictly on the basis of the aetiological model. In any case, function attributions will inevitably suffer from vagueness, but that shouldn't worry us. Similar problems of systematic temporal vagueness affect other areas of biology. There is no moment in time, for example, when we could observe the start of a new species. For by the time some single freak of nature has become the first common ancestor of a species, it and all potential observers will be long dead.
19. The importance of such acquisitions of function in evolution has become well known thanks to S. J. Gould's concepts of spandrels and exaptations. (This use of the word *spandrel*, in fact, is a cultural exaptation.) See Gould and Lewontin (1979).
20. See Bigelow and Pargetter (1987), Proust (1997), and for discussion Walsh and Ariew (1996).
21. Not all treatments of social constructionism fit this caricature. For a particularly balanced and enlightening treatment, see Hacking (1999).

Chapter 3

1. See Ameisen (2004), who refers to the 2002 Nobel Prize–winning work of Robert Horvitz on *C Elegans*.

2. For a rich trove of information about this thriving area, including Java-animated illustrations of genetic algorithms, cellular automata, and ant navigation (both of which will come up in a moment), see the splendid site set up by Jean-Philippe Rennard at http://www.rennard.org//alife/english/entree.html. The classic work on the application of ant models to more general problems in computation and artificial intelligence is Bonabeau, Dorigo, and Théraulaz (1999).
3. The metaphor of a "fitness landscape," which supports the reasoning above, does not enjoy universal assent. Sergey Gavrilets has suggested that this metaphor may at least partly be an artifact of the usual way of illustrating the problem in three dimensions, with two dimensions representing the relative frequencies of two rival alleles and the third representing fitness. In practice, a complete fitness landscape for an organism would require two dimensions for every base pair in the organism's genome, so the landscape turns out to be far more complicated than a three-dimensional representation can handle. Gavrilets opts for a metaphor of a "holey" landscape, rather than a "hilly" one. See Gavrilets (2004, chap. 5). That is, the high dimensionality of the fitness landscape makes it so that there are multiple paths between the various high-fitness genotypes. As a result, far from high-fitness genotypes being isolated "peaks" separated by low fitness "valleys," high-fitness genotypes tend to form connected networks interrupted here and there by low-fitness "holes." Local adaptation, then, consists in climbing out of a "hole" in the fitness landscape to a higher-fitness genotype. This high-dimensional representation makes the dangers of local maxima far less confining. Intelligent decisions may have their equivalent of holes in the unforeseeable long-term consequences of good and bad luck, and the notorious impossibility of labeling most events as definitively good or bad before those distant consequences can be observed.
4. See Sober (1998). The following pages draw and comment on this article.
5. In chapter 5, we will see that, in the light of "Newcomb's problem," there is reason to doubt that it is always rational to choose the act that "dominates" in the sense intended here. If we give up the

dominance principle, the distinction upon which Sober builds his argument becomes moot.
6. The canonical statement of Bayesian decision theory is Jeffrey (1965). More about it in chapter 5 below.
7. The English language has a clear way of marking that crucial distinction, but unfortunately many writers now seem oblivious to it. Molloy *might* have been a contender, meaning that it was once possible for him to become one. After a particular event has occurred, its nonoccurrence is no longer possible; yet it *might not have happened*, meaning that at one time it wasn't yet fated to happen. Moreover, if I happen to be ignorant of the fact, then *it may not have occurred* is still true *for all I know*. In the idiolect of many sloppy writers, there is no difference between "Al Gore may have won the 2000 election" and "Al Gore might have won the 2000 election." But there is a highly significant distinction: the former, but not the latter, entails that we don't actually know that he didn't.
8. In Saul Kripke's modal semantics, a "rigid designator," such as a proper name, tracks an individual across the possible worlds in which he exists (Kripke 1980). In David Lewis's equally influential semantics, a "counterpart relation" replaces cross-world identity. Identity is not strictly conserved in other possible worlds (Lewis 1986). The distinction between general and singular possibility can be made out in either system, but Kripke's semantics makes it more natural to speak of accessibility of worlds as *relative to an individual* referred to by a rigid designator.
9. If you have no one to play this game with, you can now play it in any of seventeen languages with a surprisingly ingenious machine at http://www.20Q.net. On one recent run, the machine correctly guessed *rat* on the basis of the following fifteen questions, each of which adds a specification and eliminates some potential candidates: 1. Animal. 2. Mammal. 3. Does not involve contact with other humans. 4. Not strictly vegetarian. 5. Smooth. 6. Has no spots. 7. Has claws. 8. Sometimes grey. 9. Doesn't grunt. 10. Does not weigh more than 1 tonne. 11. Can be lifted. 12. Does not live in the ocean. 13. Can be kept in a cage. 14. Sometimes found on a farm. 15. Has a tail.

10. One can strive to make up tricky counterexamples, such as "The only thing designated by the present expression"; "The being than which none greater can be conceived"; or simply "The tallest woman now living." But in all those expressions, although language strains to provide a guarantee of unique existence, it must fail. In the first two examples, it cannot warrant existence; in the third, it cannot warrant uniqueness. Note also that the third example cheats by implicitly specifying a particular, the human species. Our species should be regarded as an extended particular, defined in terms of a particular, the earth. For nothing that failed to belong to a lineage of earth-born organisms would count as a woman.
11. Ann MacKenzie was the first to point out, in unpublished work in the 1970s, that machines such as pop dispensers were quasi-intentional in the sense described.
12. A short story by Jorge Luis Borges (1998) attests to the conceivability of the alternative by imagining a society that considers it a bizarre eccentricity to believe that objects subsist and can be reidentified after they have disappeared from view.
13. There is an extensive literature on the Wason test. See, for example, Evans, Newstead, and Byrge (1993). In what follows, I follow Cosmides and Tooby (1992, 181–84).
14. Jerry Fodor has denied this, on the ground that in cheater detection, the subject is charged with conditionally enforcing an imperative, whereas the card-turning version requires verification of a conditional statement (Fodor 2000, 101–4). Checking on q just in case p is not precisely the same task as checking the truth of *if p then q*, in that the former task requires you to do nothing unless p is true. But that accounts only for one typical mistake, which consists in checking only the p card. Many subjects also unnecessarily check the $\sim q$ card. Consensus on a clear diagnosis is still lacking; the difficulty of accessing and applying the purely abstract schema, however, is well established.
15. Howard Gardner has distinguished at least seven kinds of intelligence, weakly or not at all correlated. He writes: "Specific capacities, ranging from the perception of the angle of a line to the production

of a particular linguistic sound, are linked to specific neural networks. From this perspective, it makes much more sense to think of the brain as harboring an indefinite number of intellectual capacities, whose relationship to one another needs to be clarified" (Gardner 1999, 20). More informally, have we not all occasionally observed in our close friend the surprising concurrence of shining intelligence with stubborn stupidity?

16. Cheng (1986), cited by Carruthers (2002).
17. After writing this, I became aware of a fascinating paper by David Velleman, which elaborates a somewhat similar thought about the Genesis story of the Fall. According to Velleman, what results from eating the fruit of knowledge and occasions shame is the sudden awareness of the possibility of resisting our instinctual impulses:

> "On this interpretation, the reason why Adam and Eve weren't ashamed of their nakedness at first is not that their anatomy was perfectly subordinate to the will but rather that they didn't have an effective will to which their anatomy could be insubordinate. In acquiring the idea of making choices contrary to the demands of their instincts, however, they would have gained, not only the effective capacity to make those choices, but also the realization that their bodies might obey their instincts instead, thus proving insubordinate to their newly activated will. Hence the knowledge that would have activated their will could also have opened their eyes to the possibility of that bodily recalcitrance which Augustine identified as the occasion of their shame." (Velleman 2001, 34)

Chapter 4

1. Quoted in Canguilhem (1994, 330).
2. Gould (2002, 595–743). Ghiselin (1974) was the first to defend this point of view, which has since received extended discussion. See de Sousa (1989) and, for a more recent summary, Ereshevsky (2002).
3. These discoveries have led to an entire movement that is strongly critical of the "genocentric" perspective adopted here. See Oyama

([1985] 2000), Moss (2003), and, above all, Gould (2002), whose argument I examine in the pages that follow.

4. Smith ([1776] 1937). The full text of *The Wealth of Nations* is available of several Web sites, including that of the Adam Smith Institute at http://www.adamsmith.org/smith/won-intro.htm.

5. See chapter 2 above. A number of Web sites present the structures engendered by different base rules in cellular automata. Particularly noteworthy is Steve Wolfram's site at http://mathworld.wolfram.com/ElementaryCellularAutomaton.html, which defines the most elementary CA (in a single dimension) in which each cell has only two neighbors. Because the local environment of a cell (itself included) contains only three cells, of which each has two possible states, there are eight possible configurations. There are thus $2^8 = 256$ possible rules that could determine the subsequent state of a cell. Eighty-eight are fundamentally nonequivalent. The above-mentioned site presents the patterns that result from a number of applications of each of these rules.

6. The literature of so-called creation science consistently falls back on this argument. See Pennock (2001), which contains an excellent sampling of articles for and against the version of creationism called Intelligent Design theory.

7. According to Ronald Fisher's fundamental law, "The rate of increase in fitness of any organism at any time is equal to its genetic variance in fitness at that time" (Edwards 1994). Among the drawbacks of sex, apart from the expense of feeding unproductive males, is the "cost of meiosis," which is, for any gene in a chromosome, that it stands only half a chance of making it into the successful gamete. Furthermore, the genome of any existing and viable phenotype is broken up, in sexual reproduction, to make an entirely novel and untried genotype. Such a bold policy of extreme diversification, implemented at every occurrence of sexual reproduction, while promoting the delights of diversity all around us, is obviously risky. See Maynard Smith (1978).

8. This argument was adumbrated by Darwin but given its rigorous formulation by Fisher (1930). Another elegant illustration of the logic of ESS is the "hawks and doves" scenario devised by Maynard Smith

(1984) and popularized by Dawkins (1976). "Hawks" are ready to fight to the death. "Doves" are pacific and will retreat from any confrontation; thus they will always lose out to hawks, but they are never seriously hurt. When almost all members of the population are hawks, doves have the advantage, since hawks will often be seriously wounded or killed in combat. As soon as doves become numerous enough, however, hawks regain the upper hand. We can therefore expect an oscillation about an ideal equilibrium point, which is the ESS. Note that the ESS does not necessarily refer to a ratio of *individual* doves to hawks. A single individual could play either role. The equilibrium holds over events, not individuals. It reflects the number of *occasions* on which some individual adopts a "dove" strategy, in relation to the number of occasions in which some individual adopts a "hawk" strategy.

9. The qualifier is necessary because anthropologists take issue with the biological concept of kinship as having no necessary relation to the notion of "anthropologically significant" kinship, which varies considerably from one society to another (Sahlins 1976). To be sure, the anthropological notion of kinship is far more complex than the simple arithmetic of genetic kinship. This extended conception of kinship plays a central role in the life of groups and the individuals that compose it. However, from the biological point of view, we can still treat these subtleties as deviations from the standard teleology that regulates kin selection. It would be interesting to investigate the question of whether these deviations would at some time have had consequences for group selection. As far as I know, this has not been done.

10. The expression in quotation marks was supposedly scribbled on a napkin around 1930 by J. B. S. Haldane. It is quoted in Depew and Weber (2003, 230). The mathematical details were worked out by Hamilton (1963).

11. For an excellent discussion of this paradox, including its application to evolution, see Malinas and Bigelow (2004).

12. See Henrich and Boyd (1998), Boyd and Richerson (1992), and Fehr and Fischbacher (2003).

Chapter 5

1. This image, taken from Lewis Carroll, is often used to illustrate the arms race between parasites and their hosts. See Ridley (1993).
2. See Searle (1969). The idea originated with Anscombe (1976).
3. Both are needed: neither suffices alone. To maximize the number of true beliefs would be easy, since that would be trivially achieved by believing everything. But that is equivalent to having no opinion. Similarly, universal skepticism would be too facile a hedge against false belief. Less evident is the extent of the difficulties that this duality engenders, the problem being that the means of attaining these two respective goals conflict. Tighten your defenses against falsehood, and you will miss some truths. Be more welcoming to the chance of acquiring a true belief, and you risk believing more falsehoods. This is why epistemology is irredeemably complicated.
4. The term *absolute belief* might be misunderstood. It excludes degrees of truth or of confidence, specifically those that can be assimilated to subjective probability, which I shall come to shortly. All the same, an absolute belief can vary indefinitely in other dimensions: it can be more or less firm or capricious or well founded.
5. Remember that what defines a "fair" wager is precisely that its expected value is equivalent to that of *not gambling at all*. I will return to this point on page [147] below.
6. For the original version, see Pascal ([1660] 1995, III §233 ff.), also available online at http://www.leaderu.com/cyber/books/pensees/pensees.html.
7. In truth, Pascal's argument is fallacious for another reason. The alternatives that it presents are not exhaustive. In devoting himself to the faith prescribed by one sect, Pascal risks the damnation promised for infidels by all the others. I ignore this detail, as it is not relevant to the present discussion.
8. If what one wants is not just health and happiness for oneself, but to lead a better life, one should take note of a recent study that has found "the strongly theistic, anti-evolution south and mid-west having markedly worse homicide, mortality, STD, youth pregnancy,

marital and related problems than the northeast where societal conditions, secularization, and acceptance of evolution approach European norms." As George Monbiot (2005) concluded, "If you want people to behave as Christians advocate, you should tell them that God does not exist."

9. See Tomasello and Call (1997), Hauser (2000), and Proust (2003).
10. Nussbaum (1978, §7, 701a18–20).
11. Experimental research seems to confirm that a Bayesian representation accurately accounts for the dynamic processes implicated in the exercise of a competence refined by practice, such as hitting a ball in tennis, which depends on the interaction of an acquired knowledge and sensory feedback. See Körding and Wolpert (2004).
12. See Kahneman, Slovic, and Tversky (1982) and Kahneman and Tversky (2000) for the first point of view; for the second, see Gigerenzer, Todd, and ABC Research Group (1999) and Gigerenzer and Selten (2001).
13. Plato, in *Theaetetus,* was already conscious of this dilemma, which reflects the duality of epistemic goals—believe what is true, avoid error—mentioned above. He noted that our mental "tablets" of memory could be hard or soft. Those that are easily engraved are also too easily erased, while others are hard to make a dent on, but equally hard to erase.
14. An excellent summary and an indispensable list of references is in Stich (2000). On specifically emotional irrationality, see Elster (1999), and de Sousa (2003).
15. See also Slovic (1999), who shows that our attitudes to risk are generally too pessimistic, that they depend on a configuration of ideological and political opinions that bear no logical relation to the risk in question, and that we have a tendency to confound *risk* with *ineffectiveness,* and *absence of risk* with *effectiveness.*
16. A similar duality can be found in the domain of morals. Supporters of *consequentialism*, which enjoins us to choose the action that will lead to the best possible outcome, are divided into two camps. "Act utilitarians" want to judge each action according to its probable

consequences, while "rule utilitarians" invoke the criterion of consequences only to justify a general *rule*, which in turn will determine right behavior in any particular case.

17. See (Hughes 1980); (Lee 1984); (Philips 1989).
18. See Quine (1960) and Davidson ([1993] 2001).
19. Nozick (1993, 16).
20. For some classic discussions of the Frame Problem, see Pylyshyn (1987).
21. I have elaborated this thesis in greater detail in de Sousa (1987). For confirmation from neurology, see Damasio (1994; 1999).
22. This idea, which goes back at least to the work of Scheler ([1954] 1973), has recently become the object of a great variety of interpretations and commentaries. See, among others, de Sousa (1987; 2001), Gibbard (1990), Mulligan (1998), Tappolet (2000), and Livet (2002).
23. Descartes ([1649] 1984, §74). Far from having committed the "error" attributed to him by Damasio (1994), Descartes asserted the corporeal nature of the emotions, and anticipated the discovery of their critical role in rationality in the work of Damasio and others.

References

Ainslie, George. 2001. *Breakdown of Will*. Cambridge: Cambridge University Press.

Ainslie, George, and John R. Monterosso. 2003. "Building Blocks of Self-Control: Increased Tolerance for Delay with Bundled Rewards." *Journal of the Experimental Analysis of Behavior* 89:37–48.

Allen, Colin, Marc Bekoff, and George Lauder, eds. 1998. *Nature's Purposes: Analyses of Function and Design in Biology*. Cambridge, Mass: MIT Press.

Ameisen, Jean-Claude. 2004. "Looking for Death at the Core of Life in the Light of Evolution." *Cell Death and Differentiation* 11(1):2–10.

Amundson, Ron, and George V. Lauder. 1994. "Function Without Purpose: The Uses of Causal Role Function in Evolutionary Biology." *Biology and Philosophy* 9:443–69.

Anscombe, G. E. M. 1976. *Intention*. 2nd ed. Ithaca, N.Y.: Cornell University Press.

Aristotle. 1984. "Rhetoric." In *The Complete Works of Aristotle: The Revised Oxford Translation*, ed. Jonathan Barnes, 2152–2269. Bollingen Series. Princeton, N.J.: Princeton University Press.

Arrow, Kenneth Joseph. 1964. *Social Choice and Individual Values*. 2nd ed. New York: Wiley.

Austin, J. L. [1956] 1970. "A Plea for Excuses." In *Philosophical Papers*. 2nd ed. Ed. J. O. Urmson and G. J. Warnock, 175–205. London: Oxford University Press.

Axelrod, Robert. 1984. *The Evolution of Cooperation*. New York: Basic Books.

Behe, Michael J. 1998. *Darwin's Black Box: The Biochemical Challenge to Evolution*. New York: Simon & Schuster.

Bergson, Henri. 1944. *Creative Evolution*. Trans. Arthur Mitchell, foreword by Irwin Edman. New York: Modern Library.

Bigelow, J., and R. Pargetter. 1987. "Functions." *Journal of Philosophy* 84:181–96.

Bohm, David. 1957. *Causality and Chance in Modern Physics*. Princeton, N.J.: Van Nostrand.

Bonabeau, Eric, Marco Dorigo, and Guy Théraulaz. 1999. *Swarm Intelligence: From Natural to Artificial Systems*. New York: Oxford University Press.

Boorse, Christopher. 1976. "Wright on Functions." *Philosophical Review* 85:70–86.

Borges, Jorge L. 1998. "Tlön, Uqbar, Orbis Tertius." In *Collected Fictions*, trans. Andrew Hurley, 68–81. New York: Penguin.

Boyd, Robert, and Peter J. Richerson. 1992. "Punishment Allows the Evolution of Cooperation (or Anything Else) in Sizable Groups." *Ethology and Sociobiology* 13:171–95.

Boyer, Pascal. 2001. *Religion Explained: The Evolutionary Origins of Religious Thought*. New York: Basic Books.

Bratman, Michael. 1987. *Intention, Plans, and Practical Reason*. Cambridge, Mass.: Harvard University Press.

Buss, Leo W. 1987. *The Evolution of Individuality*. Princeton, N.J.: Princeton University Press.

Canguilhem, Georges. [1966] 1994. "Le Tout et la Partie dans la Pensée Biologique." In *Etudes d'Histoire et de Philosophie Des Sciences Concernant les Vivants et la Vie*. Libraire philosophique. Paris: J. Vrin.

Carroll, Lewis. [1893] 1982. "Sylvie and Bruno Concluded." In *The Complete Illustrated Works*. New York: Gramercy.

Carroll, Sean. 2005. *Endless Forms Most Beautiful: The New Science of Evodevo and the Making of the Animal Kingdom*. New York: Norton.

Carruthers, Peter. 2002a. "The Cognitive Functions of Language." *Behavioral and Brain Sciences* 25(6):657–74; 705–10.

———. 2002b. "The Roots of Scientific Reasoning: Infancy, Modularity, and the Art of Tracking." In *The Cognitive Basis of Science*, ed. Peter Carruthers, Stephen Stich, and Michael Siegal, 73–97. Cambridge: Cambridge University Press.

———. 2003. "The Mind Is a System of Modules Shaped by Natural Selection." In *Contemporary Debates in the Philosophy of Science*, ed. Christopher Hitchcock, 293–11. Oxford: Blackwell.

Cheng, K. 1986. "A Purely Geometric Module in the Rat's Spatial Representation." *Cognition* 23(2):149–78.

Cherniak, Christopher. 1986. *Minimal Rationality*. Cambridge, Mass.: MIT Press.

Chisholm, Roderick M. 1957. *Perceiving: A Philosophical Study*. Ithaca, N.Y.: Cornell University Press.

Churchland, Paul M. 1979. *Scientific Realism and the Plasticity of Mind*. Cambridge: Cambridge University Press.

Clifford, William K. 1886. "The Ethics of Belief." In *The Ethics of Belief and Other Essays*, 70–98. Amherst, N.Y.: Prometheus (1999). Available online at http://ajburger.homestead.com/files/book.htm.

Condorcet, Marie Jean Antoine Nicolas Caritat, Marquis de. [1785] 1973. *Essai sur l'Application de l'Analyse à la Probabilité des Décisions Rendues à la Pluralité des Voix*. Facsimile Reprint of 1785 edition. Bronx, N.Y.: Chelsea.

Cosmides, Leda, and John Tooby. 1992. "Cognitive Adaptations for Social Exchange." In *The Adapted Mind: Evolutionary Psychology and the Generation of Culture*, ed. Jerome H. Barkow, Leda Cosmides, and John Tooby, 163–228. New York: Oxford University Press.

Cowie, Fiona. 1999. *What's Within: Nativism Reconsidered*. New York: Oxford University Press.

Cummins, Robert. 1975. "Functional Analysis." *Journal of Philosophy* 72:741–64.

Daly, Martin, and Margo Wilson. 2003. "Do Pretty Women Inspire Men to Discount the Future?" *Royal Society Proceedings B* 10.1098/rsbl.2003.0134. Biology Letters.

Damasio, Antonio R. 1994. *Descartes' Error*. New York: J. P. Putnam's Sons.

———. 1999. *The Feeling of What Happens: Body and Emotion in the Making of Consciousness*. New York: Harcourt Brace.

Davidson, Donald. [1967] 1980. "Causal Relations." In *Essays on Actions and Events*, 149–62. Oxford: Clarendon Press.

———. [1993] 2001. "Radical Interpretation." In *Inquiries into Truth and Interpretation*, 125–39. Oxford: Clarendon Press.

Dawkins, Richard. 1976. *The Selfish Gene*. New York: Oxford University Press.

de Sousa, Ronald. 1987. *The Rationality of Emotion*. Cambridge, MA: MIT Press.

———. 1989. "Kinds of Kinds: Individuality and Biological Species." *International Studies in the Philosophy of Science* 3(2):119–35.

———. 2000. "Deux Maximes de Rationalité Émotive." In *Emotion und Vernunft; Émotion et Rationalité*. Sudia Philosophical vol. 59, ed. Emil Angehrn and Bernard Baertschi, 15–32. Bern; Stuttgart; Wien: Paul Haupt.

———. 2001. "Émotions Morales." In *Dictionnaire de l'Éthique et de Philosophie Morale*, 3ᵉ ed., dir. Monique Canto-Sperber, 619–28. Paris: Presses Universitaires de France.

———. 2003. "Paradoxical Emotions." In *Weakness of Will and Practical Irrationality*, ed. Sarah Stroud and Christine Tappolet, 274–97. New York: Oxford University Press.

DeNault, Lisa K., and Donald A. MacFarlane. 1994. "Reciprocal Altruism Between Male Vampire Bats, Desmondus Rotundus." *Animal Behavior* 49:855–56.

Depew, David J., and Bruce H. Weber. 2003. *Darwinism Evolving: Systems Dynamics and the Genealogy of Natural Selection*. Cambridge, MA: MIT Press.

Descartes, René. [1641] 1986. *Meditations on First Philosophy with Selections from the Objections and Replies*. Trans. John Cottingham, Robert Stoothoff, and Dugald Murdoch. Cambridge: Cambridge University Press.

———. [1649] 1984. *The Passions of the Soul*. Vol. 1 of *The Philosophical Writings of Descartes*. Trans. John Cottingham, Robert Stoothoff, and Dugald Murdoch. Cambridge: Cambridge University Press.

Dunbar, Robin I. M. 2003. "The Social Brain: Mind, Language, and Society in Evolutionary Perspective." *Annual Review of Anthropology* 32:163–81.

Edelman, Gerard. 1987. *Neural Darwinism: The Theory of Neuronal Group Selection*. New York: Basic Books.

Edwards, A.W. F. 1994. "The Fundamental Theorem of Natural Selection." *Biological Reviews* 69:443–74.

Elster, Jon. 1999. *Alchemies of the Mind: Rationality and the Emotions*. Cambridge; New York: Cambridge University Press.

Ereshevsky, Marc. 2002. "Species." *Stanford Encyclopedia of Philosophy*. (Fall 2002 Edition); Edward N. Zalta (ed.); http://plato.stanford.edu/archives/fall2002/entries/species/.

Evans, Jonathan St. B. T, Stephen E. Newstead, and Ruth M.J. Byrge. 1993. *Human Reasoning: The Psychology of Deduction*. Hove: Lawrence Erlbaum.

Fehr, Ernst, and Urs Fischbacher. 2003. "The Nature of Human Altruism." *Nature* 425 (23 Oct.):785–92.

Fisher, R.A. 1930. *The Genetical Theory of Natural Selection*. Oxford: Clarendon Press.

Fitelson, Branden, and Ellott Sober. 2001. "Plantinga's Probability Argument Against Evolutionary Naturalism." In *Intelligent Design and Its Critics: Philosophical, Theological, and Scientific Perspectives*, ed. Robert E. Pennock, 411–27. Cambridge, Mass.: MIT Press.

Fodor, Jerry. 1983. *The Modularity of Mind*. Cambridge, Mass.: MIT Press.

———. 2000. *The Mind Doesn't Work That Way: The Scope and Limits of Computational Psychology*. Cambridge, Mass.: MIT Press.

Gardner, Howard. 1999. *Intelligence Reframed: Multiple Intelligences for the 21st Century*. New York: Basic Books.

Gardner, Martin. 1970. "The Fantastic Combinations of John Conway's New Solitaire Game 'Life.'" *Scientific American* 233 (Oct.):120–23.

Gavrilets, Sergey. 2004. *Fitness Landscapes and the Origin of Species*. Princeton, N.J.: Princeton University Press.

Gehring, Walter. 1999. *Master Control Genes in Development and Evolution: The Homeobox Story*. New Haven: Yale University Press.

Ghiselin, M. T. 1974. "A Radical Solution to the Species Problem." *Systematic Zoology* 23:536–44.

Gibbard, Allan. 1990. *Wise Choices, Apt Feelings: A Theory of Normative Judgment*. Cambridge, Mass.: Harvard University Press.

Gigerenzer, Gerd, and Reinhard Selten, eds. 2001. *Bounded Rationality*. Cambridge, Mass.: MIT Press.

Gigerenzer, Gerd, Peter Todd, and ABC Research Group. 1999. *Simple Heuristics That Make Us Smart*. New York: Oxford University Press.

Gilbert, Margaret. 1992. *On Social Facts*. Princeton, N.J.: Princeton University Press.

Gould, Stephen Jay. 2002. *The Structure of Evolutionary Theory*. Cambridge, Mass.: Belknap Press.

Gould, Stephen Jay, and Richard L. Lewontin. 1979. "The Spandrels of San Marco and the Panglossion Paradigm: A Critique of the Adaptationist Programme." *Proceedings of the Royal Society of London* B 205:581–98.

Hacking, Ian. 1999. *The Social Construction of What?* Cambridge, Mass.: Harvard University Press.

Hamilton, W. D. 1963. "The Evolution of Altruistic Behavior." *American Naturalist* 97:354–56.

Hardin, Garret. 1968. "The Tragedy of the Commons." *Science* 162:1243–48.

Hauser, Marc D. 2000. *Wild Minds: What Animals Really Think*. New York: Henry Holt.

Henrich, J., and Robert Boyd. 1998. "The Evolution of Conformist Transmission and the Emergence of Between-Group Differences." *Evolution and Human Behavior* 19:215–42.

Hofstadter, Douglas R., and Douglas McGraw 1995. "Letter Spirit: Esthetic Perception and Creative Play in the Rich Microcosm of the Roman Alphabet." In *Fluid Concepts and Creative Analogies: Computer Models of the Fundamental Mechanisms of Thought*, ed. Douglas R. Hofstadter and the Fluid Analogies Research Group, 407–66. New York: Basic Books.

Hughes, R. I. G. 1980. "Rationality and Intransitive Preferences." *Analysis* 40(3):132–43.

Hume, David. [1779] 1947. *Dialogues Concerning Natural Religion*. Ed. and intro. by Norman Kemp-Smith. Indianapolis, Ind.: Bobbs Merrill.

Jacob, François. 1982. *The Possible and the Actual*. Seattle: University of Washington Press.

Jeffrey, Richard C. 1965. *The Logic of Decision*. New York: McGraw Hill.

Kahneman, Daniel. 2000. "Evaluation by Moments: Past and Future." In *Choices, Values, and Frames*, ed. Daniel Kahneman and Amos Tversky,

693–708. Cambridge New York: Cambridge University Press Russell Sage Foundation.

——— Paul Slovic, and Amos Tversky, eds. 1982. *Judgment Under Uncertainty: Heuristics and Biases*. Cambridge: Cambridge University Press.

Kahneman, Daniel, and Amos Tversky, eds. 2000. *Choices, Values, and Frames*. Cambridge: Cambridge University Press.

Kant, Immanuel. 1998. *The Metaphysics of Morals*. Trans. Mary J. Gregor, Intro. Roger J. Sullivan. Cambridge Texts in the History of Philosophy. Cambridge: Cambridge University Press.

Kauffman, Stuart. 1995. *At Home in the Universe: The Search for the Laws of Self-Organization and Complexity*. New York: Oxford University Press.

Körding, Konrad, and Daniel M. Wolpert. 2004. "Bayesian Integration in Sensorimotor Learning." *Nature* 427/6971: 244–47.

Kripke, Saul A. 1980. *Naming and Necessity*. Cambridge, Mass.: Harvard University Press.

Laplace, Pierre-Simon, Marquis de. 1830. *The System of the World. Translated from the French, and Elucidated, with Explanatory Notes*, trans. Henry H. Harte. Dublin: Longmans, Rees, Orme, Brown, and Green.

Lee, Richard. 1984. "Preference and Transitivity." *Analysis* 44(3):129–34.

Lem, Stanislaw. 1974. "*De Impossibilitate Vitae* and *De Impossibilitate Prognoscendi*." In *A Perfect Vacuum: Perfect Reviews of Non-Existent Books*, trans. Michael Kandel, 141–66. San Diego: Harcourt Brace and Company.

Lennox, James G. 2001. *Aristotle's Philosophy of Biology*. New York: Cambridge University Press.

Lewens, Tim. 2002. "Adaptationism and Engineering." *Biology and Philosophy* 17:1–31.

Lewis, David. 1986. *On the Plurality of Worlds*. Oxford: Blackwell.

Libbrecht, Kenneth, and Patricia Rasmussen. 2003. *The Snowflake: Winter's Secret Beauty*. Stillwater, Minn.: Voyageur Press.

Livet, Pierre. 2002. *Émotions et Rationalité Morale*. Paris: Presses Universitaires de France.

Malinas, Gary, and John Bigelow. 2004. "Simpson's Paradox." In *The Stanford Encyclopedia of Philosophy*, ed. Edward N. Zalta, Spring 2004

Edition. URL = http://plato.stanford.edu/archives/spr2004/entries/paradox-simpson/.

Marr, David. 1982. *Vision*. San Francisco: Freeman.

Maynard Smith, John. 1978. *The Evolution of Sex*. Cambridge: Cambridge University Press.

———. 1984. "Game Theory and the Evolution of Behavior." *The Behavioral and Brain Sciences* 7:95–126.

Maynard Smith, John, and Eörs Szathmáry. 1999. *The Origins of Life: From the Birth of Life to the Origins of Language*. New York: Oxford University Press.

Melden, A. I. 1961. *Free Action*. London: Routledge and Kegan Paul.

Millikan, Ruth. 1984. *Language, Thought, and Other Biological Categories*. Cambridge, Mass.: MIT Press.

Millikan, Ruth Garrett. [1989] 1993. "In Defense of Proper Functions." In *White Queen Psychology and Other Essays for Alice*, 13–29. Cambridge, Mass.: MIT Press.

Monbiot, George. 2005. "Better Off Without Him." *The Guardian* (Oct. 11). Available online at http://www.monbiot.com/archives/2005/10/11/better-off-without-him/.

Monod, Jacques. 1972. *Chance and Necessity: An Essay on the Natural Philosophy of Modern Biology*. New York: Vintage Books.

Moss, Lenny. 2003. *What Genes Can't Do*. Cambridge, Mass.: MIT Press.

Mulligan, Kevin. 1998. "From Appropriate Emotions to Values." *Monist* 81(Jan. 1):161–88.

Nozick, Robert. 1993. *The Nature of Rationality*. Princeton, N.J.: Princeton University Press.

Nussbaum, Martha. 1978. *Aristotle's De Motu Animalium: Text with Translation, Commentary, and Interpretative Essays*. Princeton, N.J.: Princeton University Press.

Ogien, Ruwen, and Christine Tappolet. Forthcoming. *Repenser les Normes et les Valeurs*. Paris.

Olds-Clarke, Patricia. 1997. "Models for Male Infertility: The t Haplotypes." *Reviews of Reproduction* 2:157–64.

Oyama, Susan. [1985] 2000. *The Ontogeny of Information: Developmental Systems and Evolution*. 2nd rev. and exp. ed. Durham, N.C.: Duke University Press.

Pape, Robert A. 2003. "The Strategic Logic of Suicide Terrorism." *American Political Science Review* 97 (Aug):343–61.

Pascal, Blaise. [1660] 1995. *Pensées and Other Writings*, trans. Honor Levi. New York: Oxford University Press.

Pennock, Robert T., ed. 2001. *Intelligent Design: Creationism and Its Critics. Philosophical, Theological and Scientific Perspectives*. Cambridge, Mass.: MIT Press.

Philips, Michael. 1989. "Must Rational Preferences Be Transitive?" *Philosophical Quarterly* 39, 477–83.

Popper, Karl R. 1972. *Objective Knowledge: An Evolutionary Approach*. Oxford: Clarendon Press.

Prigogine, Ilya, and Isabelle Stengers. 1984. *Order Out of Chaos: Man's New Dialogue with Nature*. Boulder: New Science Library.

Prior, Arthur N. 1968. "Identifiable Individuals." In *Papers on Time and Tense*. Oxford: Clarendon Press.

Proust, Joelle. 1997. *Comment l'Esprit Vient Aux Bêtes: Essai sur la Représentation*. NRF Essais. Paris: Gallimard.

———. 2003. *Les Animaux Pensent-Ils?* Paris: Bayard.

Pylyshyn, Zenon, ed. 1987. *The Robot's Dilemma: The Frame Problem and Other Problems of Holism in Artificial Intelligence*. Norwood, N.J.: Ablex.

Queneau, Raymond. 1961. *Cent Mille Milliards de Poèmes*. Paris: Gallimard.

Quine, Willard Van Orman. 1960. *Word and Object*. Cambridge, Mass.: MIT Press.

Ridley, Mark. 2000. *Mendel's Demon: Gene Justice and the Complexity of Life*. London: Weidenfeld & Nicolson.

Ridley, Matt. 1993. *The Red Queen: Sex and the Evolution of Human Nature*. New York: Macmillan Viking.

Sahlins, Marshall. 1976. *The Use and Abuse of Biology: An Anthropological Critique of Sociobiology*. Ann Arbor: University of Michigan Press.

Scheler, Max. [1954] 1973. *The Nature of Sympathy*, trans. Peter Heath, introd. by W. Stark. Hamden, Conn.: Archon Books.

Searle, John R. 1969. *Speech Acts: An Essay in the Philosophy of Language*. Cambridge: Cambridge University Press.

Skinner, B. F. 1953. *Science and Human Behavior*. New York: Free Press.

Slovic, Paul. 1999. "Trust, Emotion, Sex, Politics, and Science: Surveying the Risk-Assessment Battlefield." *Risk Analysis* 19(4):689–701.

Smith, Adam. [1776] 1937. *An Inquiry into the Nature and Causes of the Wealth of Nations*, ed. and notes by Edwin Cannan. New York: Modern Library.

Sober, Elliott. 1994. "The Primacy of Truth-Telling and the Evolution of Lying." In *From a Biological Point of View: Essays in Evolutionary Biology*. Cambridge Studies in Philosophy and Biology. Cambridge: Cambridge University Press.

———. 1998. "Three Differences Between Deliberation and Evolution." In *Modeling Rationality, Morality, and Evolution*, ed. Peter A. Danielson, 408–22. New York: Oxford University Press.

Sober, Elliott, and David Sloan Wilson. 1998. *Unto Others: The Evolution and Psychology of Unselfish Behavior*. Cambridge, Mass.: Harvard University Press.

Stephens, Christopher. 2001. "When Is It Selectively Advantageous to Have True Beliefs? Sandwiching the Better Safe Than Sorry Argument." *Philosophical Studies* 105:161–89.

Stich, Stephen. 2000. "Is Man a Rational Animal?" In *Questioning Matters: An Introduction to Philosophical Analysis*, ed. D. Kolak, 221–36. Mountain View, CA: Mayfield.

Strawbridge, William J., Richard D. Cohen, Sarah J. Shema, and George A. Kaplan. 1997. "Frequent Attendance at Religious Services and Mortality over 28 Years." *American Journal of Public Health* 87(June 6):597–607.

Tappolet, Christine. 2000. *Emotions et Valeurs*. Paris: Presses Universitaires de France.

Tomasello, Michael, and Josep Call. 1997. *Primate Cognition*. New York: Oxford University Press.

Tooby, John, and Leda Cosmides. 1992. "The Psychological Foundations of Culture." In *The Adapted Mind: Evolutionary Psychology and the*

Generation of Culture, ed. Jerome H. Barkow, Leda Cosmides, and John Tooby, 19–136. New York: Oxford University Press.

Velleman, David. 2001. "The Genesis of Shame." *Philosophy and Public Affairs* 30(1):27–52.

Vogel, Steven. 1999. *Life's Devices: The Physical World of Animals and Plants.* Princeton, N.J.: Princeton University Press.

Walsh, Denis M. 2002. "Brentano's Chestnuts." In *Functions: New Essays in the Philosophy of Psychology and Biology*, ed. Robert Cummins, André Ariew, and Mark Perlman, 314–37. New York: Oxford University Press.

Walsh, Denis M., and André Ariew. 1996. "A Taxonomy of Functions." *Canadian Journal of Philosophy* 26:493–514.

Watts, Duncan. 2003. *Small Worlds: The Dynamics of Networks Between Order and Randomness.* Princeton, N.J.: Princeton University Press.

Westermarck, Edward. 1922. *The History of Human Marriage.* London: Macmillan.

Wigner, Eugene P. 1960. "The Unreasonable Effectiveness of Mathematics in the Natural Sciences." *Communications in Pure and Applied Mathematics* 13:1–14.

Wilkinson, Gerald S. 1984. "Reciprocal Food Sharing in the Vampire Bat." *Nature* 308:183.

Williams, George C. 1966. *Adaptation and Natural Selection.* Princeton, N.J.: Princeton University Press.

———. 1992. *Natural Selection: Domains, Levels, and Challenges.* New York: Oxford University Press.

Winsor, Mary P. 2003. "Non-Essentialist Methods in Pre-Darwinian Taxonomy." *Biology and Philosophy* 18:387–400.

Wolf, Arthur P. 1970. "Childhood Association and Sexual Attraction: A Further Test of the Westermarck Hypothesis." *American Anthropologist* New Series 72(3):503–15.

Wolfram, Stephen. 2003. *A New Kind of Science.* Champaign, Ill.: Wolfram Media.

Wynne-Edwards, Vero C. 1962. *Animal Dispersion in Relation to Social Behavior.* Edinburgh: Oliver and Boyd.

Index

Abraham, 31–2
adaptation, 24–6, 29, 40, 44, 62, 89, 115, 178
Agamemnon, 31–2
akrasia. *See* weakness of will
algorithms and notation, 135–6
alphabet, 16–18
alternative medicine, 132
altruism, 27, 68, 84, 89, 104, 108–16
   and Simpson selection, 111–13
   fascism as a form of, 117
   reciprocal, 109–10
animal mimicry, 105–7
animal suicide, 85
   and the legend of the lemmings, 89
ant navigation, 59–62, 113, 116, 159
aphasia, 76, 79
apoptosis, 59
Aristotle, 3, 7, 33–6, 39, 42–3, 129–30, 155, 174
Arrow, Kenneth, 103
Asimo, robot, 154
Aspectual principle, 152

bad money drives out good, 54
base rate, 134
Bayesian decision theory, 67, 130, 143, 151
belief, 77
   absolute, 123–4, 126–7, 130
   and detection, 9, 129, 130, 145
   as subjective probability, 126
   importance of, defined, 127
Bergson, 8, 49
Bernard, Claude, 88
better safe than sorry, principle, 127
Brentano, Franz, 73

Cabanis, Pierre Jean Georges, 36
Carruthers, Peter, 76, 79, 137, 146
causation
   and book-keeping, 95, 114
   and explanation, 35–40
   and teleology, 30–2. *See also* teology, aetiological account of
   Aristotelian, 34
   holds between particulars, 74
   local, 97–102
   material, 36
cellular automata, 97–101, 155, 159, 163
cheater-detection, 7–9, 136
Chisholm, Roderick, 73
Churchland, Paul, 77
Clifford, William, 128, 152
competition and cooperation, 104
computation, 4, 11, 12, 13, 15, 17

Condorcet, Marie-Jean-Antoine
    Nicolas Caritat, Marquis de,
    103
conflict. *See* values: multiplicity of
    between levels of organization,
    95
    inner, 86
    of interests vs. of values, 83, 85
connectionism, 59
consciousness, 9, 20
Conway
    John. *See* Game of Life
cooperation
    and sanctions, 114–19
critical transition in evolution or
    development, 56
culture, 46, 114–19
cybernetic model. *See* feedback

Dawkins, Richard, 91, 94, 95, 96,
    105, 109, 164, 170
Deep Blue. *See* Asimo
deliberation, 130
    and natural selection, 63–9
        Descartes, 12, 21, 22, 72, 153,
        156, 167
desire, 54
determinism, 33, 34, 155
digital and analog, 12–18, digitality,
    27
    of time, in Cellular Automata, 99
direction of fit, 121, 140
discounting
    future, 146

diversity, 4, 29, 37, 39, 46, 80, 106, 118
DNA, 18, 19, 62, 70, 71, 72, 94
    as information, 70, 93, *See also*
    replicators
drinking and flying, 157
dynamical systems, 17, *See also*
    governor, Watt's steam

economics, 53, 54, 97, 105, 151
    applies literally to biology, 53, 120
EEA. *See* environment of
    evolutionary adaptation
emergence, 27, 97
emotions, 28, 54, 96, 121, 144, 147,
    151, 152, 150–3, 167
    arbitrators of rationality, 152, 153
environment of evolutionary
    adaptation (EEA), 89, 115, 133,
    134, 146
Equus, 58
error, first and second types," 132
ESS. *See* evolutionarily stable
    strategy
essence, 34, 39, 72, 155
Euthyphro problem, 150–3
evolution, 4, 9–10, 18–19, 22, 27, 28, 33,
    41, 43, 56, 80, 84, 88, 91, 93, 100,
    101, 115, 120, 146, 148, 165, 166
    and freaks of nature, 43
    critical transitions in, 70–2, 104
    knowledge of, 81
    of mind, 20–6 evolutionarily
        stable strategy (ESS), 105, 106,
        107, 118, 164

fatalism, 33
feedback, 48
fitness, 5, 25, 43, 51, 61, 64, 88, 90, 93, 102, 105, 108, 112–4, 120–1, 127
    and Fisher's Fundamental Theorem of Natural Selection, 163
    frequency-dependent, 102, 104–8
    landscape, 61, 159
Fodor, Jerry, 76–7, 172
form and matter, 39
forms of intelligence, 79
free will, 131
functions, 9, 26, 29–30, 39–56, 76, 85–7, 91–3, 103, 115, 149, 158
    and malfunctions, 45, 51, 116
    of deplorable traits, 54
    propensionist interpretation, 52

Galileo, 12
Game of Life, 99
game theory, 53, 121
generality
    vs. specificity and particularity, 74
generality, two senses of, 73–5
genes, 18, 53, 62, 88–96, 106, 109–10, 115, 118, 120
    and genocentrism, 95, 109, 145, 162, *See also* natural selection, beneficiaries of
    for selfishness, 108
    homeobox, 18, 62
    regulatory, 62
    selfish, 94
    working against individual interests, 85–6
genetic algorithms, 59, 159
Gide, André, 56
Gigerenzer, Gerd, 135, 166, 172
God. *See* Intelligent Design
    impenetrable purposes of, 156
Gordon, Robert, 155
Gould, Stephen J., 90–6, 163
governor, Watt's steam, 13–17, 48, 137
granular universe, hypothesis of, 101
groups. *See also* natural selection, group
    defining progeny of, 114
    identity criteria for, 113

Haldane, J.B.S., 164
Hardin, Garrett, 97–8
hill climbing. *See* natural selection, limitations of
Hume, 21, 141

imitation, 116–17
improbability, 37–8, 40
incest avoidance, 115–16
individuals, 82, 90–6
    and communities, 19–20, 27, 118
    as locus of value, 82
    counterfactual identification of, 160
    reidentification of, 82–4, 110, 161
    species as, 90–1, 161

individuals (*cont.*)
    vs. populations, 43, 82, 89–90, 107, 111, 115, 118, 128, 134, 164
induction, 23
informational encapsulation. *See* modularity
intelligence, varieties of, 161–2
Intelligent Design, 4, 37, 40–1, 96, 100, 106, 163, 172
intention, 30–2, 40–1, 56, 66, 82–3, 106
intentional inexistence, 73, 74
intentionality, 20, 73–4, 82, 84, 104, 130
    and quasi-intentionality, 75
    full-fledged, 75
invisible hand, 97–8, 101
irrationality
    and criticism, 7, 11
    and emotions, 150
    systematic, 127, 131–8

Jacob, François, 58, 70
John, Gospel of, 34–5

Kahneman, Daniel, 144–5, 166
Kant, 8, 104–5
Kauffman, Stuart, 36–7, 155
kibbutzim, 115
kin selection, 109–10, 164

language, 18, 23, 27, 57, 68, 72
    and counterfactual hypotheses, 63, 68
    and intentionality, 72–6
    and metacognition, 57
    and modularity, 76–82
    and the frame problem, 149
    topic neutral, 81, 137, 146
Laplace, Pierre Simon, Marquis de, 33–4, 155
laws of nature, 7–8, 13, 43
    merging with normativity, 141
Lem, Stanislaw, 37
lottery paradox, 126–7
lying, 104–8, 128

mathematics, 23–5, 43, 141, 147
    topic neutrality of, 77
Maynard Smith, John, 19, 54, 70–2, 104–5
meiotic distortion, 94, 102
memes, 46, 54
metacognition, 57, 65–6, 84–86
metaphors from technology, 29
Millikan, Ruth, 45
miracles, 141
modularity
    and cognition, 79–81, 136–7, 146, 153
    and computational complexity, 79–80
    and emotions, 150
    in rats, 80
Monarch butterflies, 106–7
Monod, Jacques, 4, 29
Monty Hall problem, 141–3
Morgenbesser, Sidney, 103, 139, 147

Index

natural laws. *See* laws of nature
natural selection, 4–6, 26, 101, 131, *See also* selectionist systems
  and altruism, 108
  and defects, 41
  and deliberation, 20, 58, 61–9, 84, 137
  and design, 3–6, 15, 21
  and emergence, 97
  and local causation, 97–102
  and mathematics, 24–5
  and modularity, 80–1
  and probability, 48, 82
  and truth-telling, 23
  and vestigial teleology, 28, 57
  beneficiaries of, 47, 88, 91–6, 102, 109–11, 145
  group, 23, 27, 89, 91, ? 110 –15
  limitations of, 20, 37, 41, 61
  Simpson's, 111
  units of, 27, 91, 95
Newcomb's problem, 142–3
noncontradiction, 140
norms, 11, 83, 104, 122–9, 138–41
  and criticism, 7, 11, 23, 95, 100, 123, 129, 140
  epistemic, 128
  merging with laws of nature, 141
  not admitting of degrees, 122–3
  of transitive preferences, 138–9
  of truthfulness, 105
normativity, 7, 27, 31–2, 40, 45, 52, 56, 121–3
  three levels of, 140–3
Nozick, Robert, 142, 144

Ogien, Ruwen, 122, 128, 140
operant conditioning, 46, 59–60, 131
optical illusions, 77
original sin, 86, 154, 162

Paley, William, 40
parthenogenesis, 106
Pascal, Blaise, 124–5, 128, 152
  wager, 124–5
pessimism, 153
Philebus principle, 151
Plantinga, 20–5, 81, 146, 154, 172
Plato, 16, 18, 19, 120, 150, 166
pop dispenser, 74–5
Popper, Karl, 57
possibility
  and possible worlds, 70
  general and singular, 69, 70
potentialities, 34, 44
practical syllogism, 130
principle of charity, 141, 147
prisoner's dilemma, 63, 66, 98, 118, 119
  tit for tat strategy, 66
proper names vs. common nouns, 75
proteins, 18, 39, 71, 72
Proust, Joelle, 72, 166
psychosis, 32

Queneau, Raymond,
  hundred trillion sonnets, 156

random wobbling, 61–3
rationality, 6–9, 27
  and deliberation, 31
  and success, 4, 137
  and the frame problem, 150
  and time, 144–7
  axiological, 121, 133, 150–3
  categorial vs. normative, 7, 31, 56
  evolution of, 131
  global vs. individual, 137
  group, 19, 97, 102–4, 97–119
  minimal, 27, 31, 102
  relativity of, 32, 87, 90, 147, 150
  strategic vs. epistemic, 121–4, 137, 152
reasons, 3, 7, 11, 22, 77, 87, 98, 101, 127, 132, 144
  reductionism, 91
  and design, 97
idle threat of, 96
regressive induction, 66–7
reliabilism, 22–3
religion, 29, 158, *See also* psychosis
replicators, 53, 55, 94–6
representation, 9, 23, 57, 65, 73, 90–1, 121, 135–8, 148–50
  digital and analog 12–18, 48, 57, 71, 81, 99–101, 123, 138
reproduction
  asexual, 91
  of developmental systems, 94

  sexual, 91, 93, 104, 105, 163
risk, 132–4
  and phase transitions, 133

sanctions, 114–19, 140
selectionist systems, 41, 46–8, 59, 60, 101
  and local maxima, 61
sex ratio, 105–6
Simpson's paradox, 111
Skinner, B.F., 46
Smith, Adam, 97
snowflakes, 37–8
Sober, Elliott, 63, 106, 128
social Darwinism, 53
spandrels, 158
SPEEDWATCH. *See* governor, Watt's steam
Stephens, Christopher, 127
suicide terrorists and kin selection, 116
superstition, 23, 131–3, 150
surprise examination, 66
Szathmáry, Eörs, 19, 70–2, 104

Talleyrand Périgord, Charles Maurice de, 87
Tappolet, Christine, 122, 167
teleology, 10, 26, 30–7, 40–4, 49, 52–7, 80–5, 94–6, 109, 118, 146
  aetiological account of, 44–6, 50–2, 87–8
  and causation, 30–3

and fatalism, 32–5
and feedback, 48–9
and human action, 30
varieties of, 156
vestigial, 26–8, 44, 49–57, 80, 85–6, 91, 94–6, 116, 146
thoughts of animals, 72
tinkering, in technology and evolution, 58
tragedy of the common, 97–8, *See also* prisoner's dilemma
transitivity of preferences, 138–9
transitivity, norm of, 103, 104, 138, 139, 147
tropisms, 9–11, 56, 129, *See also* desire
and natural selection, 10
Twain, Mark, on faith, 126

values
and emotions, 150
conflicts of, 83
individual, 26–7, 53, 76, 86
multiplicity of, 52–4, 58, 82–6
vs. norms, 121–4
vampire bats, 83
Vatican. *See* drinking and flying
vestigial teleology *see* teleology, vestigial
vitalism, 36
Voltaire, 43
voting paradox, 103

Wason test, 78, 81, 136
weakness of will, 144
Westermarck, Edward, 115
Wilde, Oscar, 144
Williams, George C., 91, 93
Wöhler, Friedrich, 36
Wolfram, Steven, 100, 163

Yates, Andrea, 31–2, 102

Zeno of Elea, 25

5519377